FLOW ADAPTIVE SCHEMES

Flow Adaptive Schemes

DISSERTATION
Submitted in fulfilment of the requirements of
the Board of Deans of Delft University of Technology
and the Academic Board of the International Institute for Infrastructural,
Hydraulic and Environmental Engineering for the Degree of DOCTOR
to be defended in public
on Friday, 20 December 1996 at 13:30 h

by

VEDRANA KUTIJA
Master of Science in Hydraulic Engineering, IHE Delft

A.A. BALKEMA / ROTTERDAM / BROOKFIELD / 1996

This dissertation has been approved by the promoter:
Prof. dr M.B. Abbott, TUD/IHE Delft

Other committee members:
Chairman: Rector Magnificus, TU Delft
Co-chairman: Rector, IHE Delft

Prof. dr ir J.A. Battjes, Delft University of Technology, Delft
Prof. dr J.A. Cunge, Laboratoire d'Hydraulique de France, Echirolles
Prof. dr M.J. Hall, International Institute for Infrastructural, Hydraulic and Environmental Engineering
Prof. dr E.E. Rosinger, University of Pretoria, Pretoria
Prof. dr ir G.S. Stelling, Delft University of Technology, Delft

Published by
A.A. Balkema, P.O. Box 1675, 3000 BR Rotterdam, Netherlands
(Fax: +31.10.4135947)
A.A. Balkema Publishers, Old Post Road, Brookfield, VT 05036-9704, USA
(Fax: 802.276.3837)

ISBN 90 5410 405 8

Printed in the Netherlands

Contents

APPENDICES

To Chapter 2

To Chapter 3

To Chapter 4

To Chapter 5

Abstract

The current rapid development of hydroinformatic systems demands further improvements in numerical modelling, and thus in the computational hydraulics which underlies such modelling. This work is a study in the development of flow adaptive numerical schemes in computational hydraulics that are directed to enhancing modelling capabilities. The general domain of the work is unsteady nearly horizontal flow in one and two space dimensions. It follows different advances in computational hydraulics situated in different physical situations. Four particular examples have been studied; additional flow resistance due to flexible vegetation, one dimensional supercritical flow, flow in networks of channels and nearly horizontal two dimensional flow. All of the schemes thereby developed show similar features, employing intermediate results of computation in order to achieve a better performance. By adopting the point of view of cybernetics, these numerical schemes are examined as control systems. This provides a generalisation of features common to the notion of flow-adaptivity. For the purpose of characterising flow adaptive schemes, a process of deconstruction was followed, starting with the most simple example of an implicit numerical scheme.

Acknowledgements

The financial support that made this work possible was provided by IHE. In particular I should like to thank the IHE Rector Prof. Ir. W.A. Segeren and the head of the HH department Ir. J. Luijendijk for this.

I should like to express my sincere gratitude to my promoter Prof. M.B. Abbott for his support and direction throughout the duration of my study. I would also like to thank him for the rare opportunity I have had to learn so much by working with him. My work has also benefitted from inspiring discussions with Prof J.A. Cunge, Prof. E.E. Rosinger and Prof. G.S. Stelling.

I am very thankful to the members of the Hydroinformatics section and the Computer group of IHE for the amiable and creative atmosphere in which this work was done. This gratitude should also be extended to my other colleagues and friends within and outside IHE that have always been beside me.

I should like to acknowledge my appreciation to current employer, the University of Newcastle upon Tyne, especially the head of the Water Resources Engineering Group, Prof. P.E. O'Connell, for the understanding that has been shown in enabling me to finish this work.

Finally, most of all I wish to acknowledge my dear husband Jaime Miguel for the constant and magnanimous support he has given me during all these years. Without his endless patience this work would certainly never have been accomplished.

CHAPTER 1

Flow adaptive schemes

1.1 Motivation

Some thirty years ago, under the influence of the development of von Neuman-type digital computers, computational hydraulics was born as a hydraulics reformulated to suit the possibilities and requirements of the discrete, sequential and recursive processes that are the hallmarks of such digital computation. The practice of computational hydraulics necessitated a turning back to the theoretical bases of hydraulics, as represented by the conservation laws of fluid flow in both continuous and discrete forms, and the use of a wide range of numerical methods for their solution. From that time onwards, computational hydraulics has continued to advance and in the late eighties it represented a core around which the subject area of hydroinformatics could be advanced and structured.

During all these years, new developments in related fields, such as fluid dynamics and, especially, numerical analysis, were infiltrating extensively into computational hydraulics. The first developments started with explicit finite difference schemes, which however often suffered from severe time step limitations due to numerical stability reasons. In the late sixties these were largely replaced by implicit finite difference schemes which, although they required a somewhat bigger computational effort per time step, did not have limitations on the time step used, being unconditionally stable. These implicit schemes are nowadays widely accepted and used in numerous commercial packages for the solution of most practical hydraulics problems.

However, following the trend of development of information technology generally, the numerical schemes used in computational hydraulic ought also to become more powerful by acquiring self-steering capabilities in order to fulfil the ever more demanding tasks confronting them. Various developments in the field of numerical methods for the solution of partial differential equations (e.g. Chakravarthy and Szema, 1989) are already being introduced in computational hydraulics, at least at the research level. By way of illustration we might mention: total variance diminishing schemes (TVD) (Garcia-Navarro *et al*, 1994), essentially non-oscillatory schemes (ENO) (Nujić, 1995) and adaptive grid methods e.g. (Rahman and Chaudry, 1995).

The present work can be regarded as a study, within this framework, of the possibilities of developing *flow adaptive schemes* in computational hydraulics. The term *adaptive* is nowadays widely used, and, still, it is not easy to define it. The definition adopted from numerical analysis (Rheinboldt, 1983) may however serve to introduce the notion:

> *An algorithm is termed adaptive if it uses intermediate results to modify the course of computation in such a way that results are in some sense optimal.*

The adaptation of the course of a computation is usually done through the intervention of a *feedback function*. A function which uses intermediate results as input which influences the further computation is usually chosen as a feedback function. The above-given definition of an adaptive algorithm is quite vague, and indeed it can at the same time almost serve to define a control function. This observation leads to the notion that an algorithm with some kind of built-in control structure can also be seen as an adaptive algorithm.

This approach, which up to now was never fully generalised, is opening up the field to a conceptually new generation of flow-modelling systems. In as much as it can be so generalised and uses intelligent adaptation, it aims at a fifth generation of modelling.

The term 'flow adaptive' itself derives from the observation that the usual result of a computational-hydraulics simulation is a flow field and the adaptation of the scheme consequent upon the intermediate results is then based on the resulting computed flow field.

In order to make the notion of flow adaptive schemes more clear, this Chapter will advance gradually through their various aspects. For this purpose, a quite simple numerical scheme, the 'box' scheme, will be applied to the pure advection equation. Control and adaptivity will then be introduced in such a way as to give another point of view upon numerical schemes and the numerical solutions of physical problems. It will as well provide a means to classify adaptive algorithms on a more general, because more abstract, level.

The field of adaptive numerical methods will be briefly introduced in order to emphasise the differences and possible similarities between these methods and those of the present work. Finally in this chapter, the discussion of adaptivity in computational hydraulics may serve to open the door onto this topic and to provide a short overview of what follows in this work. The possibilities for the introduction of adaptivity or control in computational hydraulic algorithms are very wide and have so far been but little exploited. In this work, four adaptive algorithms in computational hydraulics are investigated; this by no means exhausts the possibilities inherent in these methods, but indeed serves only to introduce them.

1.2 The deconstruction of the numerical scheme

The notion of deconstruction was reintroduced into French, as *déconstruction*, by Jacques Derrida (e.g 1991, p.273) in the late 1960s and has since become one of the most widely employed of all philosophical terms, having now been carried over into many languages, including English. It conveys the idea of a disassembling or disarranging the parts of a whole that leads to a better understanding of that whole. It usually suggests a certain attention to structures, whereby these structures are undone, decomposed or desedimented. What it conveys is a process that exposes how an ensemble is constituted with the aim of reconstructing this ensemble to some positive purpose.

The present work must open with a deconstruction of the numerical scheme because its subject is a certain loss of rigidity of structure in numerical schemes such as allows a restructuring. Through this process, these schemes become more flexible, more physically meaningful, more efficient and altogether able to model nature more closely.

When we write a 'meaningful' numerical scheme we make a particular kind of statement about our representation of the natural world. The 'particularity' of this statement means that a meaningful scheme of computation is in harmony, or is consistent, with our ways of perceiving and comprehending this natural world. For our present purposes of demonstrating the principles it will suffice to consider initially only the most simple of numerical schemes. We shall consider the pure advection equation:

$$\frac{\partial \phi}{\partial t} + u \frac{\partial \phi}{\partial x} = 0$$

This equation expresses non-diffusive transport of the function (matter) $\phi\,(x,t)$ with velocity u. The existence of a solution asks for one upstream boundary condition such as $\phi\,(0,t)$ and one initial condition such as $\phi\,(x,0)$.

We shall consider the box scheme that is consistent with this equation,

$$\frac{\alpha(\phi^{n+1}-\phi^{n})_{j} + (1-\alpha)(\phi^{n+1}-\phi^{n})_{j+1}}{\Delta t} + u\,\frac{\beta(\phi_{j+1}-\phi_{j})^{n} + (1-\beta)(\phi_{j+1}-\phi_{j})^{n+1}}{\Delta x} = 0 \qquad (1.1)$$

where, conventionally, $0 \leq \alpha \leq 1$, $0 \leq \beta \leq 1$, with $\alpha, \beta \in R^1$.

Now it is well known that the second-order consistent form of this equation is obtained when $\alpha = 0.5$ and $\beta = 0.5$ and that in this case it is unconditionally stable in the now classical, von Neumann-Richtmyer, sense. It is equally well known (e.g. Abbott and Basco, 1989, pp. 151-156) that this particular weighting of the scheme gives rise to basic problems in its algorithmic structuring, closely corresponding to problems of ill-posedness of the equivalent continuum problem.

In fact for the left-to-right sweep, being the proper algorithmic structure in the continuum sense when the flow velocity is from left to right, equation (1.1) takes the form:

$$\phi_j^{n+1} = \frac{((1-\alpha) - Cr\cdot\beta)\,\phi_j^n + (-\alpha + Cr(1-\beta))\,\phi_{j-1}^{n+1} + (\alpha + Cr\cdot\beta)\,\phi_{j-1}^n}{(1-\alpha) + Cr(1-\beta)} \tag{1.2}$$

with $Cr = u\,\Delta t/\Delta x$.

For the right-to-left sweep, being the proper algorithmic structure in the continuum sense when the flow velocity is from right to left, equation (1.1) takes the form:

$$\phi_j^{n+1} = \frac{((1-\alpha) + Cr\cdot\beta)\,\phi_j^n + (-\alpha - Cr(1-\beta))\,\phi_{j+1}^{n+1} + (\alpha - Cr\cdot\beta)\,\phi_{j+1}^n}{(1-\alpha) - Cr(1-\beta)} \tag{1.3}$$

For example, when a sweep in the one direction is made while the velocity is in the other direction, the scheme amplifies the determining boundary conditions in a completely unrealistic way, and indeed, when the corresponding Courant number passes through a unit magnitude in this case, the scheme experiences a division by zero at the first grid point that it encounters. This is not, of course, an instability in the above, classical, sense since it does not occur in time and the unboundedness of the operator is not associated with any process of proceeding to a limit in space and time. In the borderline case that arises when the velocity is zero, on the other hand, although no amplification occurs, a solution imposed on the boundary will still send oscillations into the solution with a wave length of $2\Delta x$, as exemplified for initial conditions of $\phi = 0\;\forall\, x$ in Fig.1.1. This 'solution' is, of course, completely unrealistic.

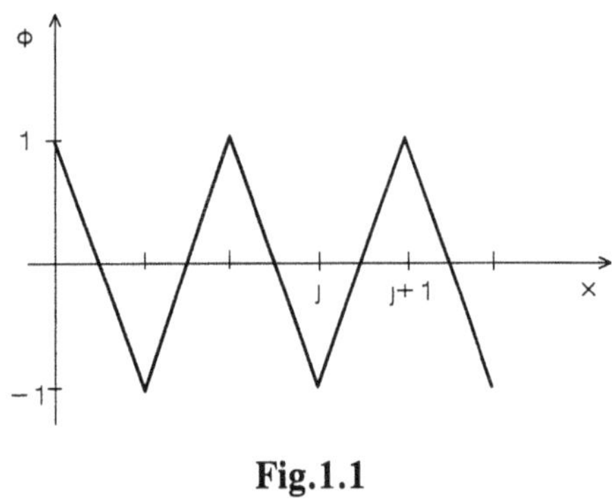

Fig.1.1

These properties of the scheme are easily corrected, however, as soon as we allow its parameters to be functions of the flow velocity u, so that the scheme becomes flow-adaptive.

Let us define a function *s(u)* as:

$$s(u) = \begin{cases} 1, & u > 0 \\ 0, & u = 0 \\ -1, & u < 0 \end{cases} \tag{1.4}$$

With the help of this function, a flow adaptive scheme for the solution of the pure advection equation can be defined as follows:

$$\frac{\alpha(\phi^{n+1}-\phi^n)_{j-s(u)}+(1-\alpha)(\phi^{n+1}-\phi^n)_j}{\Delta t}+u\cdot s(u)\frac{\beta(\phi_j-\phi_{j-s(u)})^n+(1-\beta)(\phi_j-\phi_{j-s(u)})^{n+1}}{\Delta x}=0 \tag{1.5}$$

This then leads to a unique equation that accounts for the algorithmic structure:

$$\phi_j^{n+1}=\frac{((1-\alpha)-Cr\cdot s(u)\cdot\beta)\phi_j^n+(-\alpha+Cr\cdot s(u)(1-\beta))\phi_{j-s(u)}^{n+1}+(\alpha+Cr\cdot s(u)\cdot\beta)\phi_{j-s(u)}^n}{(1-\alpha)+Cr\cdot s(u)(1-\beta)} \tag{1.6}$$

whereby, the algorithmic structure is then determined from the value of the function *s(u)* as:

$s(u) = 1$ ➡ left-to-right sweep,
$s(u) = -1$ ➡ right-to-left sweep.

For cases where the velocity is in the positive direction ($u > 0$), or opposite to that ($u < 0$), the expression (1.6) provides the form corresponding to those obtained from the classical box scheme with the appropriate algorithmic structure (equations (1.2) and (1.3)). In the case when the velocity is equal to zero ($u = 0$), equation (1.6) provides a form which gives no place to the earlier-described oscillations introduced by the sweep algorithm proceeding in the physical space x of the solution - and indeed there is then no space step involved in the algorithm at all.

This simple flow adaptive scheme can be more fully developed again through the definition of its coefficients α and β as functions of flow velocity (i.e. Courant number) so as to improve its accuracy. In order to assess the accuracy of the scheme, a truncation error analysis can be performed. Each of the terms of the equation (1.5) is then expanded into its Taylor series expansion up to third degree about the centre of the scheme. (This analysis was performed with *Mathematica*.) The truncation error so obtained has the form:

$$\text{T.E.} = \left(Cr(\beta - 0.5) + (\alpha - 0.5)\right)\frac{\partial^2\phi}{\partial x^2} + \left(Cr^2\beta(1-\beta) + \alpha(\alpha - 1) + \frac{1}{6} - \frac{Cr^2}{6}\right)\frac{\partial^3\phi}{\partial x^3} + \text{H.O.T.}$$

As observed earlier, it is commonly assumed that the highest accuracy of the scheme, in Taylor's series terms, is obtained by the use of a completely-centred scheme (with $\alpha = 0.5$ and $\beta = 0.5$) as this then leads to a second order of consistency: the first term in the truncation error vanishes. In order to achieve an even more accurate scheme in this sense, additional terms are commonly introduced into the original equation in order to cancel-out the second term in the above expression (see Abbott and Basco, 1989, p.155). As already noted, however, the second order consistent, or 'accurate', scheme can become exceeding unrealistic, not to speak of inaccurate, under certain flow conditions.

The methods of flow adaptivity proceed in another direction, in that the coefficients α and β are seen as functions of the Courant number, albeit defined in such a way to minimize the truncation error. In the first instance, the first two terms of the truncation error were set to be equal to zero and the space of solutions of this system of two nonlinear equations with two unknowns (α and β) is searched to provide

```
Solve[{Cr(β-1/2)+(α-1/2)==0,Cr^2 β(1-β)+ α(α-1)+1/6-Cr^2/6==0},{α,β}]

{}
```

As this output from *Mathematica* shows, there is no general solution to this system of equations. However, if a particular solution is requested, *Mathematica* provides two of these:

```
Reduce[{Cr(β-1/2)+(α-1/2)==0,Cr^2 β (1-β)+ α(α-1)+1/6-Cr^2/6==0},{α,β}]

α == 1 - β && Cr == 1 || α == 1 - β && Cr == -1
```

These two particular solutions show us that for a Courant number equal to unity or minus unity, by setting coefficients α and β to 0.5, an (at least) third-order accurate scheme (now in the polynomial sense) is obtained. For the other values of Courant number, however, this approach does not seem to suffice, so that another approach, based more on physical insight (such as is provided by the method of characteristics) is needed. This development passes outside the scope of the present work and will, therefore, be omitted. In the subsequent analysis in this section only the simple flow adaptive scheme given by equation (1.4) will be considered.

For the present purpose of expressing more of the structure of these schemes, we cast them into a Reverse Polish Notation, or RPN, or prefix, representation. The equation (1.2) resulting from the simple box scheme for the left - to - right sweep then appears as:

$$= \phi_j^{n+1} / + \times \phi_j^n - - 1 \alpha \times Cr \beta + \times \phi_{j-1}^{n+1} + - \alpha \times Cr - 1 \beta \times \phi_{j-1}^n + \alpha \times Cr \beta + - 1 \alpha \times Cr - 1 \beta$$

The structure of this scheme can then be exhibited as a directed graph, as shown in Fig.1.2.

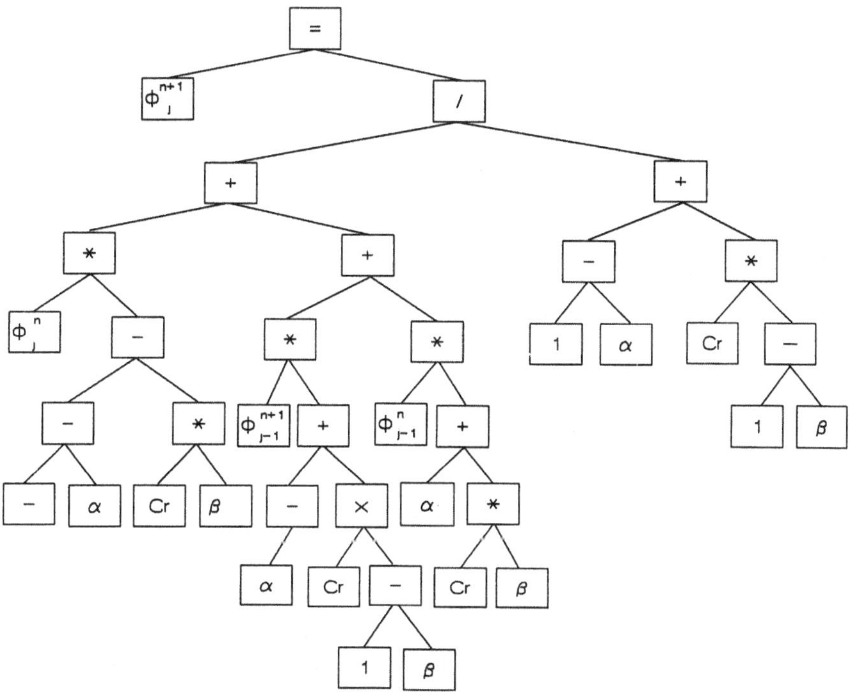

Fig.1.2

The equation (1.3) resulting from the box scheme for the right to left sweep is then cast into a somewhat different equation:

$$= \phi_j^{n+1} / + \times \phi_j^n + - 1\, \alpha \times Cr \beta + \times \phi_{j+1}^{n+1} - - \alpha \times Cr - 1 \beta \times \phi_{j+1}^n - \alpha \times Cr \beta - - 1\, \alpha \times Cr - 1 \beta$$

which leads to a different (but homomorphic) graph representation, as shown in Fig.1.3.

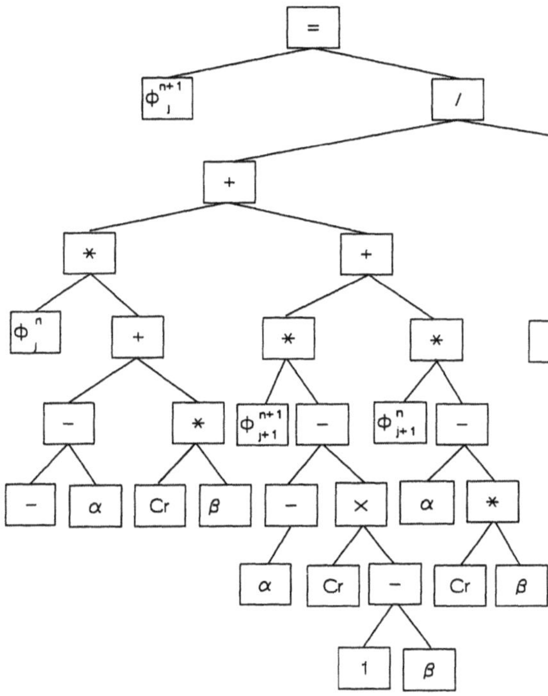

Fig.1.3

As already pointed out, the application of the simple box scheme to the solution of the pure advection equation leads to two different forms, represented by equations (1.2) and (1.3) or their graphical presentations in Fig.1.2 and Fig.1.3, dependant on the algorithmic structure. Moreover, in the case of no flow (zero flow velocity) neither of the two is applicable. In this respect, therefore, a flow adaptive scheme, introduced earlier, offers a more complete solution.

It may now be seen, however, that equation (1.6), representing the flow adaptive scheme, can be cast into a unique *graph* that takes account of the algorithmic structure:

$$= \phi_j^{n+1} / + \times \phi_j^n - - 1\, \alpha \times \times Cr s(u) \beta + \times \phi_{j-s(u)}^{n+1} +$$

$$+ - \alpha \times \times Cr s(u) - 1 \beta \times \phi_{j-s(u)}^n + \alpha \times \times Cr s(u) \beta + - 1\, \alpha \times \times Cr s(u) - 1 \beta$$

as represented by the directed graph given in Fig.1.4.

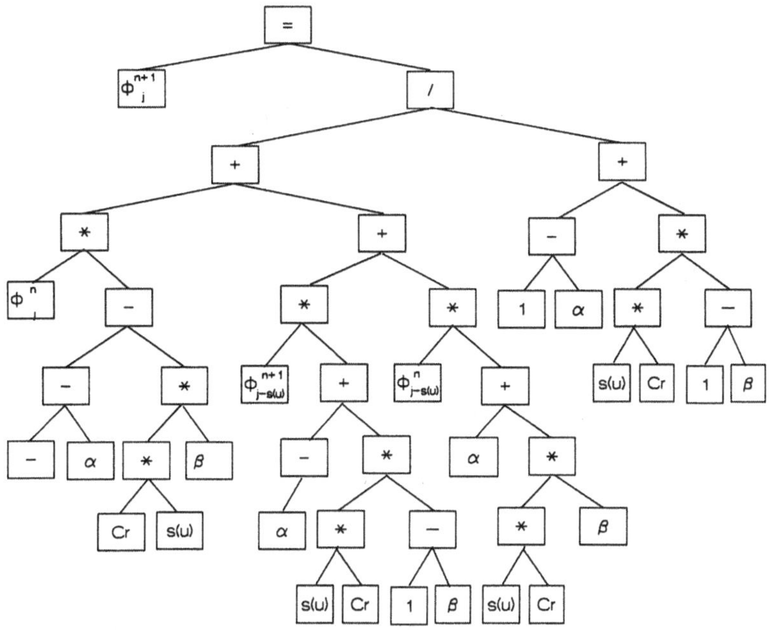

Fig.1.4

This scheme, on the other hand, contains another 'hidden' layer of calculations, consisting of the determination of the value of *s(u)* according to the function *s* defined in equation (1.4) and of the index *j-s(u)* which defines the direction of the calculation. These two calculations can as well be presented separately by directed graphs, as shown in Fig.1.5 and Fig.1.6.

In order to build up the complete picture of the flow adaptive scheme defined by equation (1.5), therefore, these two graphs should be superimposed on the graph presented in Fig.1.4. In this way an additional layer is introduced. The flow adaptive scheme is thus a scheme with another level of complexity which cannot be quit so easily represented by such directed graphs. Indeed, the graph of Fig.1.4 still does not reveal all the calculations performed in such a scheme. On the other hand, these graphs can be used to visualise the essential equivalence of schemes, as a comparison of Figs 1.2 and 1.3 illustrates, and this recognition process can itself be automated.

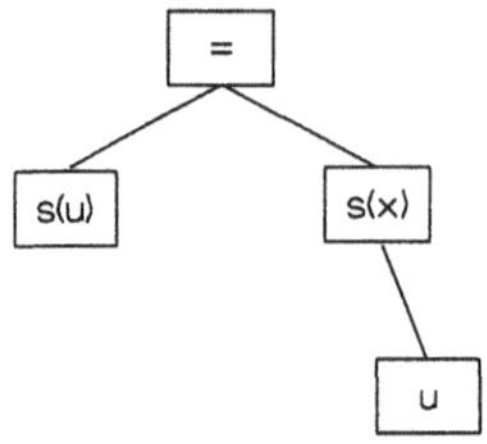

Fig.1.5

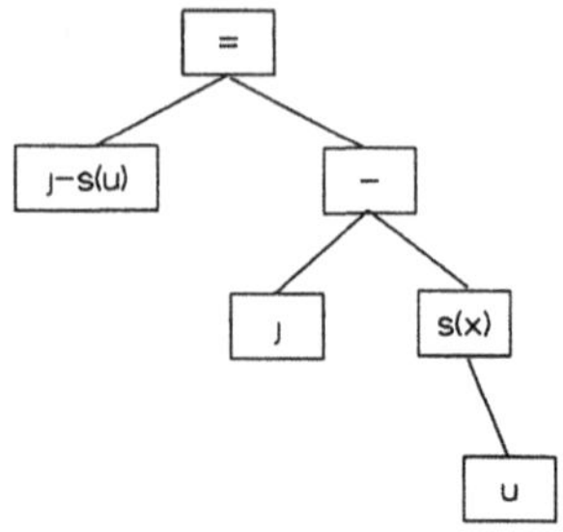

Fig.1.6

If a convention for representation of if-then-else statements in the graph form is adopted the flow adaptive scheme can, however, be represented by the directed graph of Fig.1.7

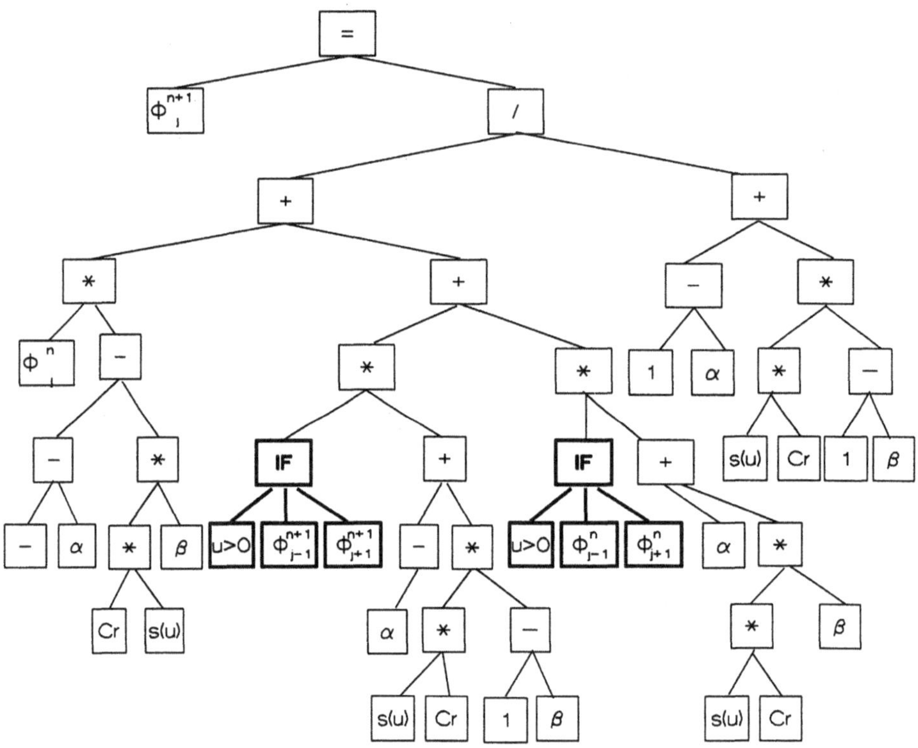

Fig.1.7

We shall not in fact make much use of such graph-theoretic methods in this connection in the rest of this work; this is not so much because of the limitations of these methods, as such, but because of the shear amount of space which their representations occupy, as exemplified, even for the quite simple scheme considered here, in Fig.1.4 and Fig.1.7.

We can still however conceive of a flow adaptive scheme as a superposition of one operator, which controls the adaptation process, upon another operator, which is the scheme itself. In this first example, the superimposed control operator is composed simply of two elementary 'if - then - else' operators so that it is essentially a *logical* operator (see Fig.1.7). The 'if' part is seen to refer to a message from the flow field which is matched against an already stored reference, so that it in fact constitutes an extremely simple *classifier*. The 'then' part feeds back to the scheme if a match is found and otherwise, through the 'else' part, it continues to the next classifier. In this example, of course, these classifiers are trivially simple, being composed of one element (or 'bit', or maybe 'trit'!) from a trinary alphabet { 0 , +1 , -1 }, but, just as clearly, this process could be extended indefinitely by allowing longer and longer strings of 'bit-like' elements in the usual way of classifier systems. Thus, even though the control operator is restricted to logical processes, it can still be extended to comprehend situations of arbitrary complexity. Further to this, we recall that, just as classifier systems can be evolved through the application of machine-learning algorithms, such as Q-learning, our control operators can be learned through analogous means (Masood and Wilson, 1995; Wilson, 1996). Although these possibilities will not be much exploited in the present work, reference will occasionally be made to them.

1.3 Control and adaptivity

The domain of the solution of a hydrodynamic problem is composed from physical space and time. For one-dimensional hydrodynamic problems this is 'the x-t plane'. However, the domain of our numerical solution is just a set of discrete points in this space. The solution of the numerical-hydrodynamic problem is then the set of values associated with these points.

We can however look at this from a different perspective. Let us first, and just for the moment, strip away the time dimension. Then, for the one-dimensional case we have a domain consisting of points distributed along the line while for the two-dimensional case we have a set of points in the plane. At each of the points of our domain we can, however, have one or more variable(s) for which we wish to solve. Now let us call the set of all these variables a *physical system representation* (PSR). The values of the variables can then be called *states of the variables* and the set of all of them can be regarded as the *state of the physical system representation.*

Let us now bring the time back into consideration. Under the action of the numerical scheme, which provides an advance in time over one time step, the physical system representation takes another state. This change of state is a *transformation* in which the numerical scheme acts as an *operator* while the physical system representation corresponds to the *operand.* (In numerical analysis it is very common to speak specifically of a numerical operator.)

A system which can be defined by a set of variables and a set of closed single-valued transformations of these variables which are dependent on the input *parameters* is in *cybernetics* called a *machine with input,* or a *transducer* (Ashby, 1973, pp.42-44). Thus a physical system representation (PSR) is a transducer. We have defined our PSR as a set of solutions of a numerical-hydrodynamic problem at any one instant of time. However, a PSR can be any set of variables describing a physical effect at a certain time, given only that a set of closed single valued transformations dependant on the input parameters is provided.

The numerical scheme can itself, as well, be seen as a transducer, with its coefficients being the set of its variables that take different values depending on the input parameters. From the point of view of cybernetics, these two transducers, being our PSR and the numerical scheme, are equivalent under certain conditions. From the point of view of their representation in a digital machines, this equivalence holds just as well. It is only for us that they can be distinguished according to the meaning that we assign to them: the variables of a PSR have for us a physical meaning while the variables of a numerical transducer are, even for us, 'just numbers'. Hence, in the rest of this work reference will be made either to the general transducer - which can be of any of the above-mentioned types - or to some other such type, or to a specific transducer called, indifferently, a PSR or a numerical operator (NO). To distinguish further between entities from each of these types, either the physical meaning we assign to the variables or the name of the numerical variables will be given.

Transducers can be coupled so that they can form new transducers. Through the coupling of our present two transducers - the one being the physical system representation and the other being the numerical operator - we have obtained a new transducer representing our dynamic system (our numerical solution of the hydrodynamic problem). The *kinematic graph* of this new dynamic system is given in Fig.1.8, where the one box stands for the one transducer and the other box for the other. The idea behind using a kinematic graph is to show more clearly the relations inside the more complex structures. Therefore, we shall use these devices (which are specific kinds of 'sign vehicles' in the language of the theory of semiotics) extensively in our descriptions of our adaptive schemes. It should be remembered, however, that the state of the physical system representation is in our case the solution of the hydrodynamic problem at a certain time, it being the set of values of the system variables at discrete points within the solution domain.

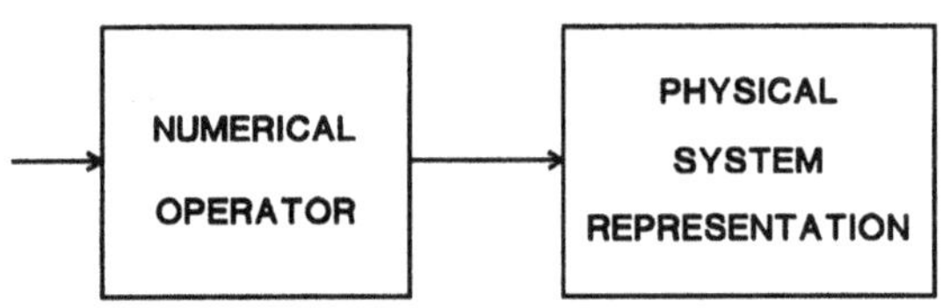

Fig.1.8

For a system of the type presented in Fig.1.8, we conventionally say that one part of it *dominates* another, as there is only one direction of influence. However, this does not represent the relations realistically in our case, for in nonlinear schemes, at least, the states of our physical system representation modify the parameters in the numerical scheme as well. When such a circularity of action exists in a dynamic system we say that there is *feedback*. A dynamic system with feedback is usually represented by a somewhat different kinematic graph, as shown in Fig.1.9.

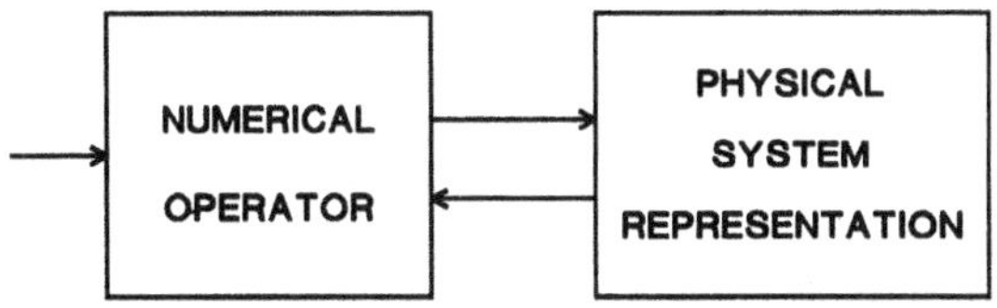

Fig.1.9

The presence of feedback in our dynamic system leads us to yet another point of view: to regard our dynamic system as a *control system*. The simplest structure of a control system consists of the *controlling* and the *controlled* parts, as shown in Fig.1.10. The similarity between the graphs shown in Fig.1.9 and Fig.1.10 then suggests that the numerical solution of a hydrodynamic problem can be treated as a controlled system. However, it should be pointed out that such a simple control system rarely exists in isolation: it usually interacts again with the surrounding medium and with other systems, forming yet more complex systems (Lerner, 1972, p.7). However, the definition of subsystems of larger systems is always possible and equally justified. It then appears that this much grosser way of viewing a scheme may provide another, and in some ways more convenient, means of following or allowing for its deconstruction then the more detailed directed-graph representation.

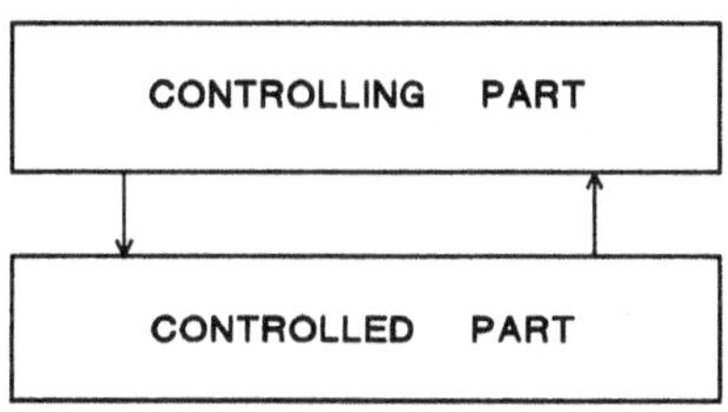

Fig.1.10

In order to avoid confining ourselves to too narrow a definition of control, let us take over the following (from Lener, 1972, p.77):

> *Control represents selected actions on some object or objects which, on the basis of the information obtained, 'improve' the functioning or development of the given object.*

The construction of a numerical-hydrodynamic solution can surely be seen as a process that is subject to such a control. The representation of our dynamic system in the classical form of a control system is shown in Fig.1.11.

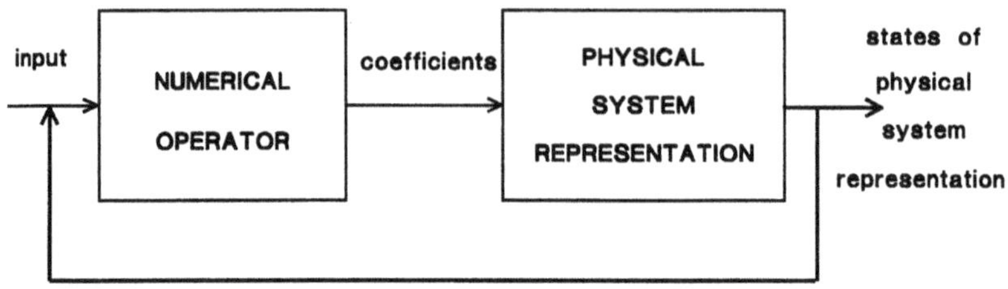

Fig.1.11

It should be stressed that until now no specific numerical method has been introduced, which means that this point of view can be maintained quite generally. Even the type of physical problem need not necessary be hydrodynamic at this stage. These considerations can, in fact, be applied to the numerical solutions of any physical phenomena which can be described by hyperbolic or parabolic partial differential equations as a 'marching' problem.

Adaptive control systems are control systems which allow one more level of feedback which then influences (by changing) some of the more essential parameters of the controller. The distinction between classical feedback and this adaptive feedback is not always very clear, however, even in the field of *adaptive control theory*. The distinction is commonly made between two time scales, whereby classical feedback is fast changing and adaptive feedback functions much more slowly (Åström and Wittenmark, 1989, p.2). When applicable, this distinction is one that will be followed throughout this work.

In fact, a numerical scheme can already be considered as an adaptive control system if this scheme is defined in the most general terms and adaptation is introduced already in the implementation phase for the tuning of the essential controller parameters. This can then be applied later, if need be, to the periodic updating of the parameters, or indeed as a 'continuous' adaptive procedure (Isermann *et al*, 1992, p.xvii).

A very simple type of adaptive control system, called *closed-loop adaptation*, is represented by the scheme shown in Fig.1.12. The adaptation loop which is the difference between the scheme of any dynamic system represented in Fig.1.11 and the scheme of an adaptive numerical scheme represented in Fig.1.12 uses the states of the PSR in much the same way as the original control loop. The only difference is that this adaptation loop does not need to act upon each and every time step but only when states of the PSR trigger it. Therefore, this loop is schematically represented in a somewhat different manner.

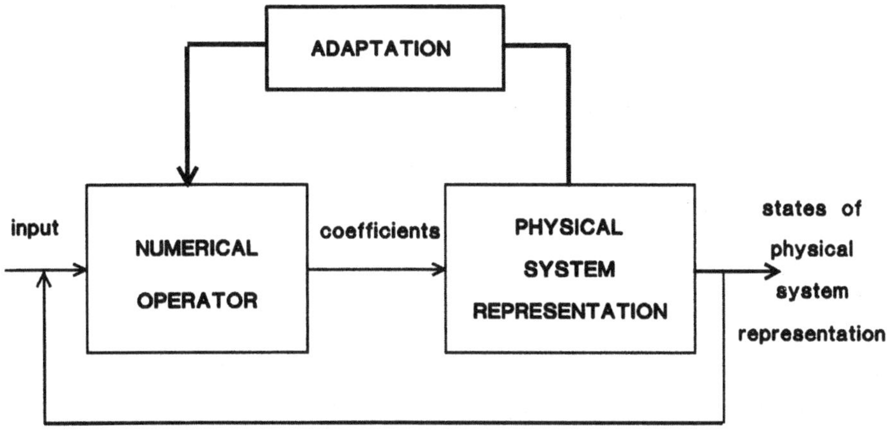

Fig.1.12

It is also possible to have an adaptive control system based on a so-called *open-loop adaptation* which does not adapt the controller on the basis of the states of the controlled system but on the basis of some external information. The architecture of such a system is shown in Fig.1.13.

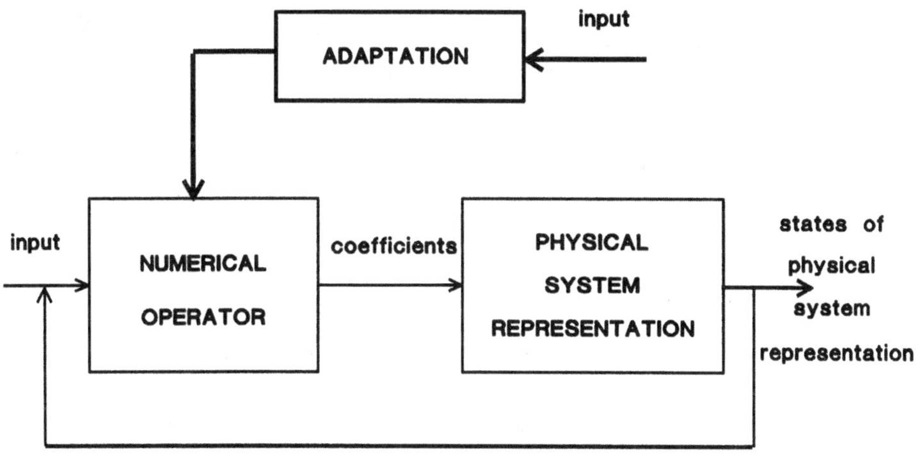

Fig.1.13

In the kinematic graphs representing the control systems of Fig.1.12 and Fig.1.13 there is a transducer named 'adaptation'. It is a source of feedback, but the fact that it is a transducer and not only a communication channel leaves a space for intermediate stages through which the information might pass before being used to transform (or adapt) the numerical operator. This transducer can of course be of any type; for example a PSR, an NO, a logical operator or a transducer composed of any these types of any required level of complexity.

1.4 Adaptive numerical methods

Within the branch of numerical analysis which deals with partial differential equations, adaptive schemes are well known. Their development was already established in the late seventies (e.g. Stepelman and Winarsky, 1979) and became widely employed in the field of boundary value problems in the form of adaptive mesh generation schemes for finite element methods (Babuška *et al,* 1983). The use of adaptive-grid methods for those initial-boundary value problems that are characterised by hyperbolic equations developed more slowly due to the problems of accuracy connected with finite difference approximations on nonuniform grids. However, at the first workshop on adaptive computational methods for partial differential equations, held in Philadelphia in 1983, there were already a few papers presented on the application of adaptive methods to solutions of hyperbolic equations (e. g. Anderson, 1983; Harten, 1983) and others again appeared in the following years (e. g. Oliger, 1984; Berger and Oliger, 1984). Most of these adaptive methods are based on the adaptation of the size of the discretisation grid. At each time step, a certain feedback function is used to update the computational grid (Rheinboldt, 1983). Therefore, even at this stage, the development of numerical grid generation was closely connected with the development of adaptive numerical methods.

The adaptivity criteria are in general divided into two main groups: those based on the assessment of error in the solution of the flow equations (so called *a posteriori* error estimates) and the more physically based criteria that take account of a specific detectable feature of the flow field, such as for example the presence of shock waves (Weatherill, 1990). From the point of view of the actual grid adaptation, two main categories of methods can be distinguished: the so called r-methods which reallocate a fixed number of grid points, and the so called h-methods which involve an automatic refinement of the grid structure (Oden, 1989). Besides all these adaptive methods concerned only with the grid size or grid distribution, the introduction of an artificial viscosity in the form of a nonlinear diffusion operator for the control of oscillations in the numerical solution can also be regarded in some ways as an adaptive feature (Oden, 1989).

Most of these methods can be considered as constituting a closed loop adaptive control system, as schematically represented by the kinematic graph of Fig.1.12. However, the adaptation transducer would represent some other combination of more simple transducers for each of the categories. The adaptation part of the methods based on *a-posteriori* error estimates would allocate one part for the error estimate and another part for the actual grid adaptation, while the other methods need just the latter one of these. The kinematic graph for the h-methods will usually be the more complicated, since the physical system representation, PSR, is itself being adapted, and not the numerical operator alone.

1.5 Adaptivity in computational hydraulics

The approach adopted in this work for the development of flow adaptive schemes differs considerably from those used in the classical adaptive numerical methods. It can still be connected with the physically-based criterion for adaptivity (the flow field), but then not with the object of adaptivity being the discretisation grid, or at least not the size of the discretisation step. The examples of the introduction of dissipative interfaces or artificial viscosity are much more in line with the kind of adaptivity studied in this work. In the case of dissipative interfaces, the change is effected directly on the results themselves. It is these that are being filtered, while the numerical scheme itself is staying unchanged. Still the effect of this adaptivity is felt in further calculations, while input parameters for the numerical operator are being changed and consequently the states of the NO are being changed as well.

As already mentioned, any numerical algorithm for the solution of hyperbolic or parabolic partial differential equations can be regarded as a control system and represented by a kinematic graph of the kind shown in Fig.1.11. If we now recall the difference between explicit and implicit finite difference schemes (e.g. Abbott and Basco, 1989, p. 234) it is clear that on the more detailed level these two classes of schemes should be represented by quite different kinematic graphs in order to grasp all of the differences in interrelations between the various parts of the control system. However, on the scale of the full time step, and with the whole domain of the problem being considered as just one PSR, they can both be represented by the scheme shown in Fig.1.12.

It will now be clear that the flow-adaptive scheme introduced earlier by equation (1.5) for the solution of the pure advection equation can as well be treated as an adaptive control system with an open-loop adaptation. Its adaptation transducer is triggered by an 'external' input, namely the flow velocity, which in this particular case is not one of the state variables of the physical system representation that is sought, as a solution, in space and time. The corresponding kinematic graph is then that given in Fig.1.13.

In each of the following examples, a different adaptive feature or parameter will either be employed in order to obtain solutions which are in some sense optimal or, in some cases, as a means of obtaining any solution at all. The possibilities for the introduction of adaptivity or control in computational hydraulic algorithms are very wide and have so far been but little exploited. In this work four adaptive algorithms in computational hydraulics are investigated; this by no means exhausts the possibilities inherent in these methods, but indeed serves only to introduce them. Each of the investigated examples is situated in a different physical domain and moreover, at least at first sight, each of them utilises a different form of adaptivity.

However, when considered as adaptive control systems and represented accordingly by their kinematic graphs, a number of similarities between them will become clear. All four of them are, for example, closed looped adaptive systems of the type schematised in Fig.1.12. The major differences between them reside in the different types of transducers used for adaptation. A short overview of the following chapters containing these examples and their form of adaptivity follows.

Chapter 2 - A numerical model for the assessment of additional resistances introduced by flexible vegetation.

In this example, the intermediate velocity profiles are used to adapt the vegetation geometry which in its turn influences the flexible resistance and consequently the velocity profile itself.

Chapter 3 - Numerical treatment of supercritical flow

In this example the intermediate results expressed in terms of the Froude number of the flow were used for guiding the adaptation of the governing equations in order to provide an approximate but exceedingly convenient solution.

Chapter 4 - A generalised solution algorithm for flow in channel networks

In this example, it is the pre-processing of the network topology that provides the control structure for the optimal management of the computational procedure.

Chapter 5 - ADI+ algorithm for the solution of two dimensional nearly horizontal flow

In this example, the intermediate flow field results are used to adapt the finite difference scheme to provide more accurate results. Although still not entirely satisfactory for practical computing, this example is introduced in order to show how the possibilities inherent in this approach can be extended to two - and potentially also three - dimensional flows.

CHAPTER 2

A numerical model for the assessment of additional resistance introduced by flexible vegetation

2.1 Introduction

The current promotion of the natural development of wetlands and other restoration projects which are directed toward more natural developments within river basins often favour the growth of reeds and other similar vegetation. Such vegetation influences the resistance of the water course and consequently the water velocities and depths. However, these influences of vegetation, and particularly flexible vegetation, on the flow resistance are still not well understood, even as a sound knowledge of these influences is necessary for effective real-time river management and other hydroinformatics applications. Consequently, laboratory experiments have been and are still being performed, while pilot river reaches with grown vegetation are being established in several regions in order to investigate this influence (Tsujimoto *et al*, 1993; Larsen *et al*, 1990; Klaassen and Urk, 1985). The numerical model described here was developed to help in deepening the understanding of vegetation-induced resistances and then, in particular, by following the effects of the various parameters involved.

This model is then used as a tool for investigating the influences of different parameters on the total resistance. It can as well, subject to minor changes, be included as a module into a two-dimensional nearly-horizontal flow model with averaged velocities for the calculation of vertical velocity profiles with flexible vegetation. This transforms the original two-dimensional model into a quasi three-dimensional model (Xian, 1993).

2.2 Adaptive structure of the algorithm

In order to capture the complexity of the interrelations between the flexible vegetation and the flow resistance, a flow adaptive approach was adopted. Based on the general adaptive control system with a single loop, as described in Chapter 1 and schematised in Fig.1.12, an adaptive algorithm was developed. This flow adaptive algorithm can best be represented by the scheme of Fig.2.1.

The numerical operator (NO) of our model is a one-dimensional vertical mixing model which is, however, coupled to the equation of conservation of momentum in the horizontal direction. The variables of the model are flow velocities at different points along the vertical. Therefore our physical system representation describes the flow field, its states being the flow velocities at different points along the domain.

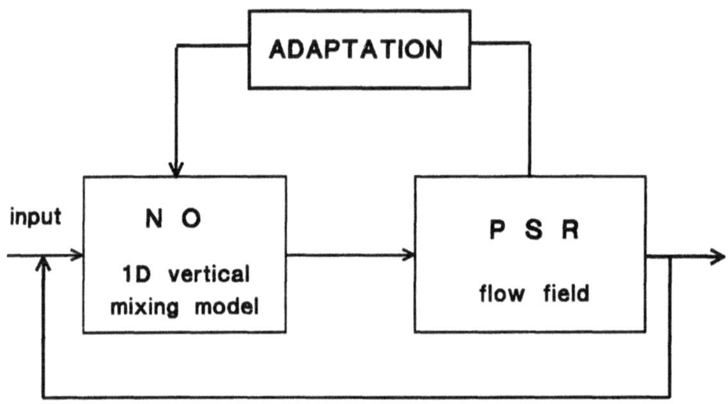

Fig.2.1

The model comprehends both rigid and flexible vegetation. For the rigid vegetation the adaptation transducer can be simple, but that applied to the flexible vegetation has to comprehend the behaviour of the rigid vegetation as well. Therefore, in the further text, only the example with the adaptive transducer for flexible vegetation will be explained. The drag force is estimated by a semi-empirical formula according to which this force is proportional to the square of the water velocity and to the projection of the vegetation in the direction perpendicular to the water flow (the *effective height of vegetation*). The bending of the flexible vegetation under the load produced by the drag force is calculated according to standard cantilever beam theory (e.g. Timoshenko, 1955). As a result of bending, the effective height of vegetation is reduced. This in turn results in a smaller drag force and less bending. The bending of the vegetation is accordingly conveniently calculated using an iterative procedure.

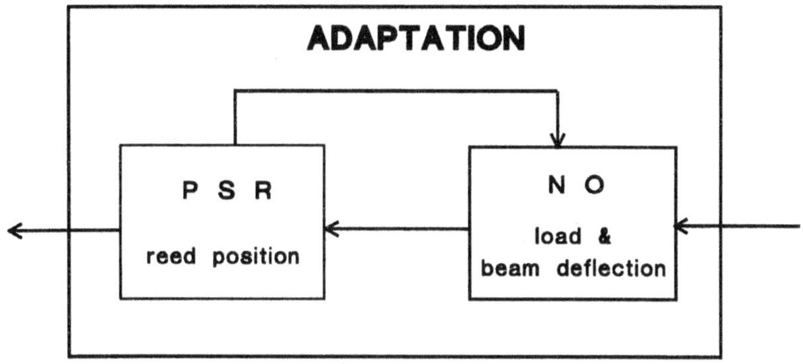

Fig.2.2

The adaptation transducer consists of another PSR, this being the reed position, and of the numerical transducer, this last being constituted by the procedure to calculate the load on the reed induced by the flow and all the parameters needed to trigger the transformation of the reed position according to the cantilever beam theory. These two transducers themselves form a control system, schematised in Fig.2.2, which works through a process of iterative convergence. Once a convergence limit is satisfied, the step out of the adaptation traducer is made and the set of parameters (state of the PSR of reed position) is transferred to the numerical operator and the calculation of the new time step can proceed.

However, it is not difficult to imagine that the PSR which describes the reed position is a part of the adaptation transducer, while reed geometry really does adapt under the influence of the flow field. If the schemes from Fig.2.1 and Fig.2.2 had been drawn together, it would become clear that this adaptive algorithm is completely symmetric. Thus, the choice of which of its parts we apportion to the adaptation and which to the original system is arbitrary. The choice was made in the present way because we are more interested in the flow field than in the reed position. It is important to point out that the same adaptive control system can be viewed from the other perspective to produce results in the form of reed positions, while the flow field would then be just a part of the adaptation transducer.

2.3 Description of the model

2.3.1 *General*

The numerical model for the assessment of the additional resistance of the flexible vegetation is based on the equation of conservation of momentum in the horizontal direction, which for the present purpose can be most conveniently written as a differential-difference equation:

$$\frac{\partial u}{\partial t} + u\frac{\partial u}{\partial x} + w\frac{\partial u}{\partial z} + \frac{1}{\rho}\frac{\partial P}{\partial x} - gi - \frac{1}{\rho}\frac{\partial \tau}{\partial z} + \frac{F_x}{\rho \Delta x \Delta z} = 0 \tag{2.1}$$

with:

u	- horizontal velocity	i	- bottom slope
w	- vertical velocity	τ	- shear stress
P	- pressure	$\frac{F_x}{\rho \Delta x \Delta z}$	- additional horizontal drag forces
ρ	- density of water		
g	- gravity acceleration		

Since this model is developed only in order to study the effects of flexible vegetation in steady uniform flow, some of the terms from the equation (2.3) are neglected. However, the model is an unsteady model used to determine steady state conditions from unknown initial conditions. Therefore, the acceleration term is not being neglected. One of the neglected terms is the pressure term. In the case that this model is used as a module within a two-dimensional nearly-horizontal flow model with averaged velocities, this term cannot of course be neglected, but it is then combined with the bottom slope term in order to provide a classical gravity term. The convective momentum terms are for the time being also neglected. Besides this, the vertical velocity w is also neglected. The reduced equation then has the form:

$$\frac{\partial u}{\partial t} - gi - \frac{1}{\rho}\frac{\partial \tau}{\partial z} + \frac{F_x}{\rho \Delta x \Delta z} = 0 \tag{2.2}$$

where $u(z,t)$, $\tau(z,t)$ and $F_x(z,t)$.

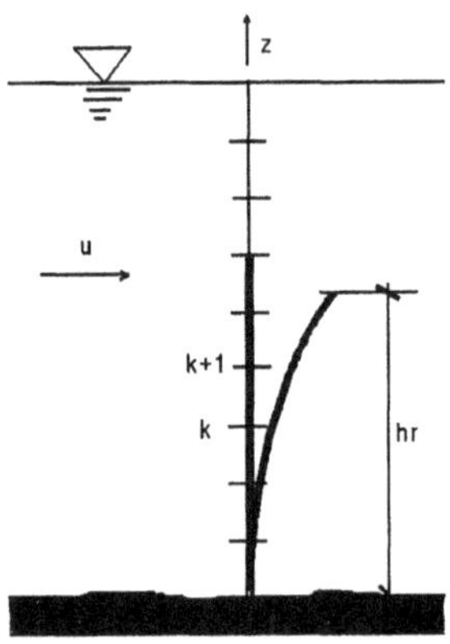

Fig.2.3

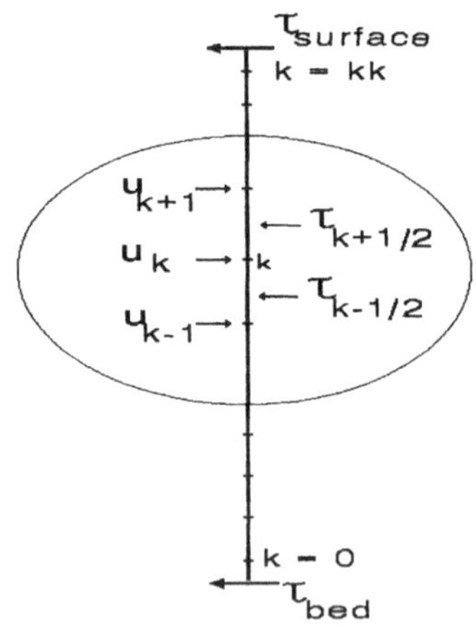

Fig.2.4

The numerical model for the solution of this equation is a one-dimensional model in the vertical direction following the schematisation of the domain as presented in the Fig.2.3 and the discretisation grid shown in Fig.2.4 with *kk* being the number of grid points. The model is based on an implicit finite difference approximation for the unknown velocities in all the discretisation points. The other two unknowns τ and F_x are expressed in terms of velocities, τ in terms of unknown velocities and F_x in terms of velocities calculated in the previous time step. The velocities are the variables of the PSR. The terms of equation (2.2) are then represented as follows.

2.3.2 *The acceleration term*

Although the model is meant for the solution for steady state conditions, it is constructed as an unsteady-state model. This feature makes possible the calculation of the unknown steady-state condition starting from any initial conditions, even though these may be unrealistic, on the basis that these initial influence will in this way be removed. This also makes it a 'marching problem' such as promotes the analogy with the cybernetic system and the development of an adaptive numerical scheme for its solution.

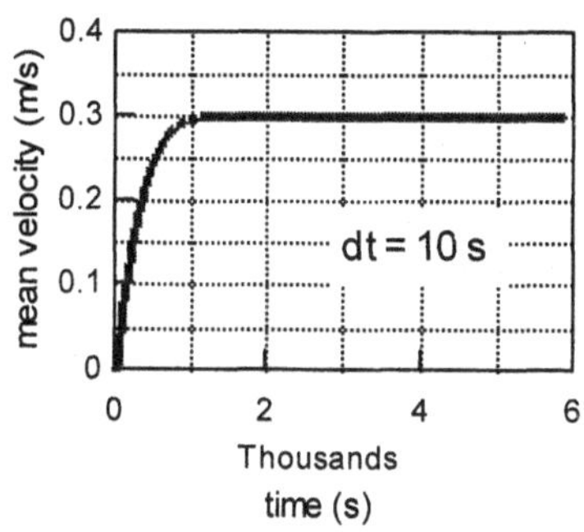

Fig.2.5

In all the examples presented in this chapter the initial velocities are equal to zero in all the grid points. The steady-state velocity profile develops gradually through the simulation time. The boundary conditions are kept constant so that with time, the steady state is approached. The mean velocity calculated as a ratio of the velocity integral over the depth and the depth itself is used to show how the steady state is approached. In Fig.2.5 the mean velocity is shown as a function of time for one of the examples. The initial velocity in the whole vertical profile was set equal to zero so that the mean velocity at t=0 was also equal to 0. It is obvious that the steady state is reached well before 2000 prototype seconds, but for security all the simulated examples were run for at least 6000 prototype seconds.

The acceleration term is approximated as follows: $\frac{\partial u}{\partial t} \sim \frac{u_k^{n+1} - u_k^n}{\Delta t}$

2.3.3 *The shear stress term*

The shear stress term is approximated within this model in two ways. In the layer of water above the vegetation it is modelled as a turbulent shear stress approximated by a mixing length theory.

$$\tau(z) = \rho \; l^2 \; \frac{\partial u}{\partial z} \left| \frac{\partial u}{\partial z} \right| \tag{2.3}$$

where l is the mixing length at the level z determined by:

$$l = \kappa z \left(1 - \frac{z}{h} \right)^{0.5} \tag{2.4}$$

with κ being the von Karman coefficient and h being the water depth.

The water velocity within the vegetated layer is rather small (Tsujimoto and Kitamura, 1990), so that a different shear stress approximation is needed in this layer. One possibility is to model this in the same way as in the upper layer using the mixing length theory, but in that case the expression for the mixing length has to be changed (Watanabe and Kondo, 1990). The other possibility is to model it according to eddy-viscosity theory (Tsujimoto and Kitamura, 1990). In the present model, the second option is implemented so that:

$$\tau(z) = \epsilon \; \frac{\partial u}{\partial z} \tag{2.5}$$

where

$$\epsilon = \rho \cdot \alpha \cdot s \cdot u$$

with α being an empirical coefficient and s being the distance between the reeds.

Comparing data from Tsujimoto with the results of the model, as will be presented later, it became obvious that the eddy-viscosity approximation was not suitable for the whole vegetation layer. Near the tip of the reed, velocities are not so small as they are deeper in the layer, and this phenomenon could not be captured with the eddy-viscosity approximation. Therefore, an additional parameter p was introduced to determine how much of the height of the vegetation layer is modelled by the eddy-viscosity approximation. The rest of the vegetation layer was then modelled in the same manner as the layer above the vegetation, following the schematisation of Fig.2.6. The influence of the parameter p on the distribution of the shear stress within the vegetation layer is considerable, as can be seen in Fig.2.7.

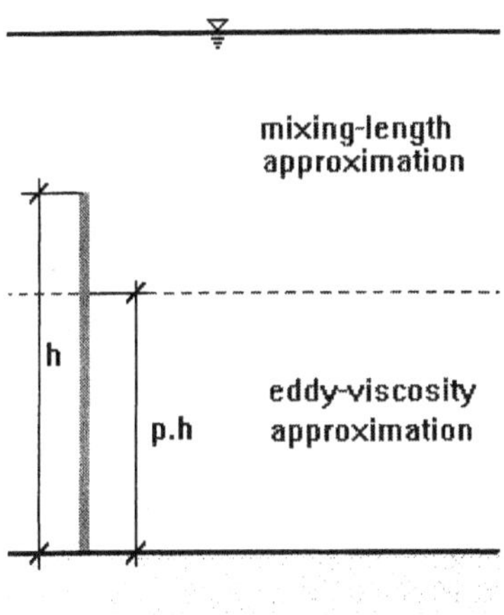

Fig.2.6

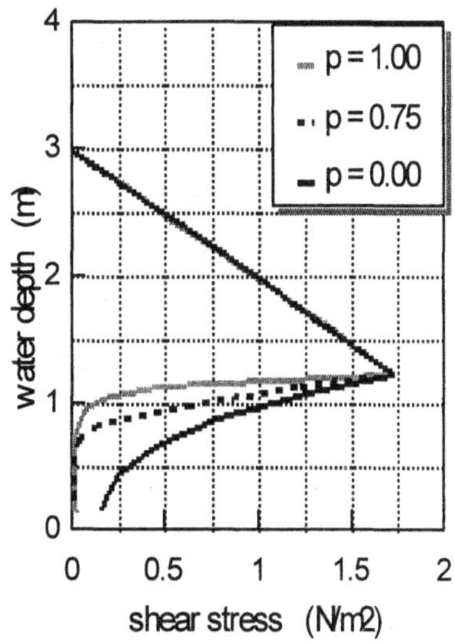

Fig.2.7

Although the parameter p was intended to function as a calibration parameter, through the investigation of the influence of various parameters it became clear that it could not in fact be treated as independent of other parameters. The density of the reeds (m), their diameter (d_r) and stiffness (EI) all in fact influence the parameter p considerably. Further investigations, not described here, are being directed to establishing a functional relationship between these parameters.

2.3.4 *The additional force term*

The additional force (F_x) is the drag force due to the vegetation. In the layer above the vegetation this force is equal to zero while in the layer with vegetation it is defined as:

$$F_x = m \frac{CD \cdot u^2 \cdot d \cdot h_r}{2} \tag{2.6}$$

with: m - density of reeds per m^2
u - flow velocity in x direction
h_r - effective hight of reed
CD - drag coefficient
d - diameter of one reed

The additional force term in equation (2.2) is then approximated as follows:

$$F_k^{n+1} = \frac{m \cdot CD}{2} \cdot d_k \cdot hr_k^n \cdot u_k^n \cdot u_k^{n+1} \tag{2.7}$$

where d_k is the diameter of the reed at the level k , hr_k^n is the portion of the reeds' effective height corresponding to the grid level k at the time level n while u_k^n and u_k^{n+1} are flow velocities at the level k at the respective times, n and $n+1$. The effective reeds' heights are in fact variables of the PSR describing the reed position. At the same time their values (states) are input parameters for the numerical transducer.

In this formulation, however, the effects of interactions between the reed induced flows are neglected.

2.3.5 *The boundary conditions*

For this model, two boundary conditions are required, one at the bed and the other at the water surface. The positions of these two boundary conditions are marked on the grid shown in Fig.2.4.

The shear stress at the bed is given through the averaged velocity and the Chezy coefficient. Although this approach is clearly not entirely satisfactory when applied to vertically nonuniform flows, it is used due to its simplicity and due to the fact that the influence of the bottom shear stress is negligible in comparison with the additional resistance of the vegetation. A potentially more accurate way would be to introduce a no-slip boundary condition, which would in practice mean that the velocity u_0 at a distance z_0(equivalent bottom roughness height) from the bottom is equal to zero (Xian, 1993).

At the water surface the slip boundary condition with no shear stress is introduced. If the combined influence of wind and the flexible vegetation is to be studied, the near-surface shear stress may be set equal to an imposed wind-induced shear stress.

The system of equations obtained upon the introduction of these finite difference approximations and the boundary conditions has as unknowns the set of horizontal velocities u_k , $k = 0,..kk$. The matrix of this system is tri-diagonal and it is solved with a double-sweep algorithm (Abbott and Basco, 1989). More detailed description of this part of the model is given in Appendix 2.1.

2.4 Bending of the reeds

The effective height of any representative reed (h_r) used in the evaluation of the drag force, is influenced by the bending of the reed due to the flow of water (See Fig.2.1). This interrelation is the main reason why the adaptive approach was adopted for development of this model. The adaptation transducer in fact consists of two transducers, one being the PSR of the reed position and the other being the NO. The bending of the reed is calculated in a separate module after each time step. This module is the adaptation transducer. As input data, the adaptation transducer uses the velocity-profile results from the previous time step. The NO uses these velocities and the states of the PSR (being the effective reed heights per grid point) to evaluate the load on the reed and to produce parameters which will trigger the transformation of the PSR. However, if the new states of PSR do differ from those used for estimation of the load for more than the prescribed limit this procedure has to be repeated. Thus, the adaptation transducer is in itself a control system as well. Details about algorithms used in this transducer are given in Appendix 2.2.

The deflection of the reed is calculated according to cantilever beam theory (e.g. Timoshenko, 1955, pp.137-165) under the load produced by the drag force. The intensity of this load is a function of the vertical position and it is by no means constant. However, in this discrete model it is approximated by segments of constant load defined by the flow velocity over each segment as schematised in Fig.2.8.

The diameter of the reed (d_r) as well as its stiffness (EI) are also not constant along the reed height. The model therefore allows the linear distribution of these values along the reed, although in the model they are again discretised in the same manner as the load.

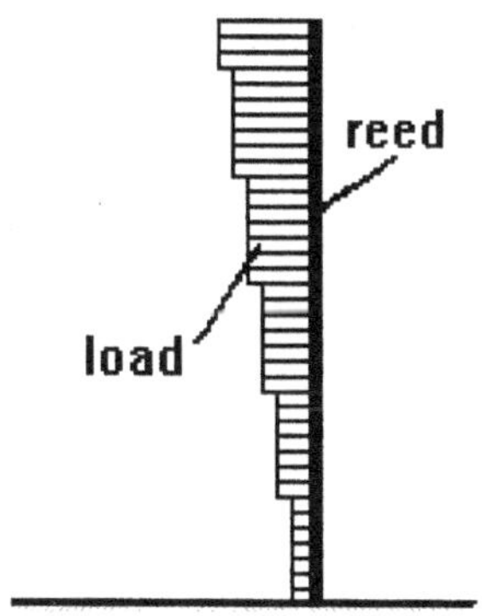

Fig.2.8

The intensity of the drag load at any level k (between points $(k\text{-}1/2)\Delta z$ and $(k+1/2)\Delta z$ is given by:

$$q_k^n = \frac{CD \cdot d_k \cdot (u_k^n)^2}{2} \tag{2.8}$$

The expression for the deflection of a point on the cantilever beam due to the partial load differs according to the position of the point relative to the position of the load. In principle, three positions can be distinguished: the point is between the fixed end of the beam and the beginning of the load; the point is within the load; and the point is between the load and the free end of the beam. Due to the discrete nature of the model, the second case was not needed. The expressions for the other two cases are as follows:

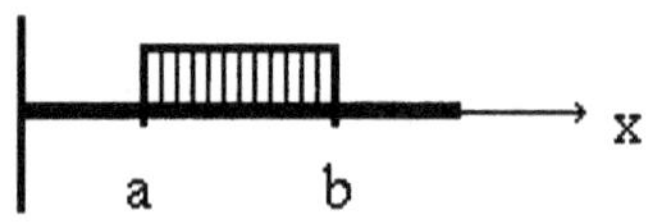

Fig.2.9

for $x \leq a$ $$\delta_x = \frac{qx^2}{12EI}\left(3b^2 - 3a^2 - 2x(b-a)\right)$$

for $x \geq b$ $$\delta_x = \frac{q}{24EI}\left(4x(b^3 - a^3) - (b^4 - a^4)\right)$$

The total deflection of the beam is constructed by superposing the deflections caused by each of the partial loads. For details see Appendix 2.2.

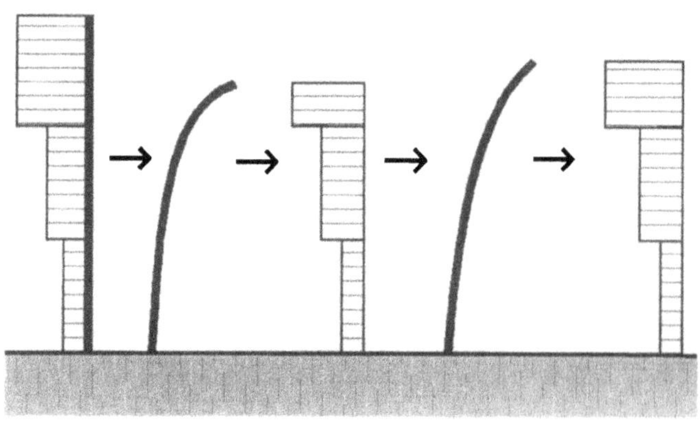

Fig.2.10

Due to bending, the effective reed height is reduced, which means that the height over which the load is acting is also reduced: this reduced load causes less bending than would result with the full effective reed height. This resulting interaction is introduced using the iterative procedure sketched in Fig.2.10. The procedure is stopped when a prescribed difference between the load height and the effective reed height is reached. This iterative procedure is in fact a form of internal control structure within the adaptation transducer. The states of the PSR influence the values of the NO which then again influence the states of the PSR.

2.5 Verification Example

Some data from Tsujimoto (Tsujimoto and Kitamura, 1990) were used for the verification of the model with fixed, inflexible, vegetation[1]; see Fig.2.11.

The parameters which were calibrated in this example were the parameter p for the definition of the layers modelled by the different turbulence models, *CD,* the drag coefficient, and α, the empirical parameter in the expression for the eddy-viscosity. The results presented in Fig.2.11 were calculated with p=0.75, CD=1.1, and α=0.005. In the following chapter, more will be said about the influence of each of these parameters.

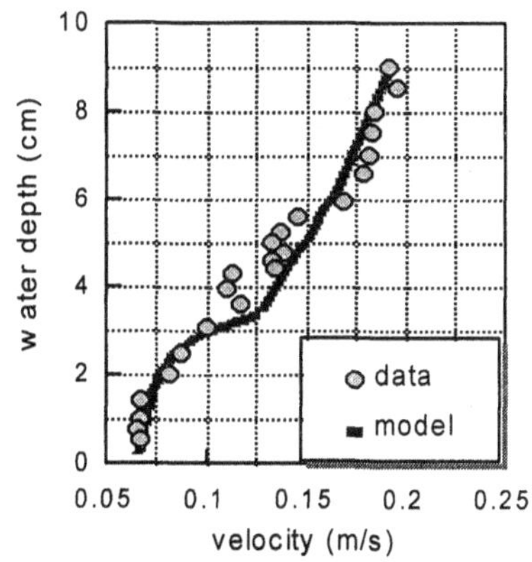

Fig.2.11

2.6 Influence of different parameters

In order to describe the influence of the different parameters on the flow resistance, a simple example is used. Its data are given below. In general, these data follow the conditions present in floods plains of the river Rhine in the Netherlands and the magnitudes of most parameters were taken from a report of Kortleve (1994).

grid step	dz	= 0.10 m	drag coefficient	CD	= 1.1
time step	dt	= 10 s	numerical parameter	p	= 0.75
bottom slope	i	= 0.0001	numerical parameter	α	= 0.01
Chezy coeff.	C	= 50 $m^{0.5}/s$	reed diameter at the bot.	dr_b	= 0.002 m
water depth	h	= 3.0 m	reed diameter at the top	dr_t	= 0.002 m
reed length	hr	= 1.25 m	reed stiffness at the bot.	Ei_b	= 2 Nm^2
reed density	m	= 100	reed stiffness at the top	Ei_t	= 2 Nm^2

[1]Unfortunately no data were available for the verification of the model with flexible vegetation.

It should be observed that the usual way of reducing the search space of relations through the introduction of dimensionless combinations of variables has not been essayed here. The aim has been to leave all such possibilities open at this stage. The influence of each of the parameters involved in the model is investigated by changing one parameter at a time from the initial set of data and comparing the results with those obtained using the initial set of data. However, not all of these parameters are independent of each other. That means that when only one parameter is changed and all the others kept constant, a situation is obtained that is not fully realistic, which might lead to premature and possibly incorrect conclusions. Therefore, mutual dependence of parameters has also to be considered when the influence of a parameter is being analysed. In the application of this model using real field data this danger should be significantly reduced due to the necessary physical consistency of data.

In the following graphs, the velocity profile obtained from the original data and the velocity profile for the case where one parameter from the initial set of data is changed are compared. The result obtained with the initial set of data is always represented by the solid black line.

2.6.1 *Computational parameters*

As could be expected, the computational parameters *dt* and *dz* do not play any significant role: see Fig.2.12 and Fig.2.13. The same conclusion holds for the total simulation time. As pointed out earlier, the simulation time for all presented results is 6000 prototype seconds in order to assure that the convergence to the steady state is realised.

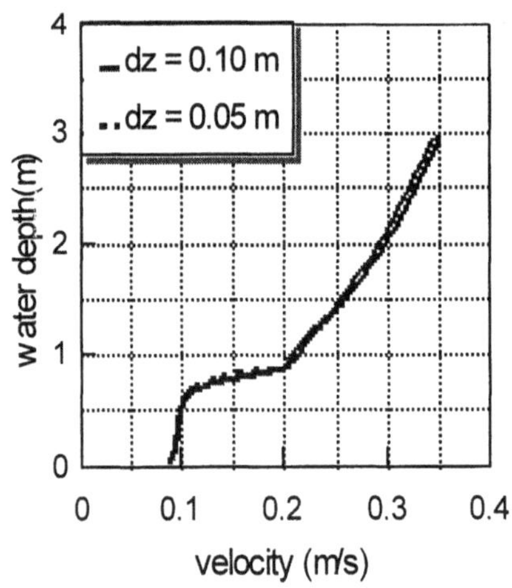

Fig.2.12

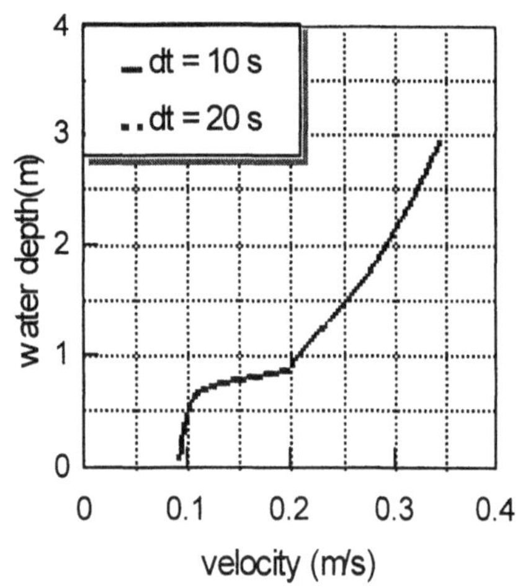

Fig.2.13

2.6.2 *Hydraulic parameters*

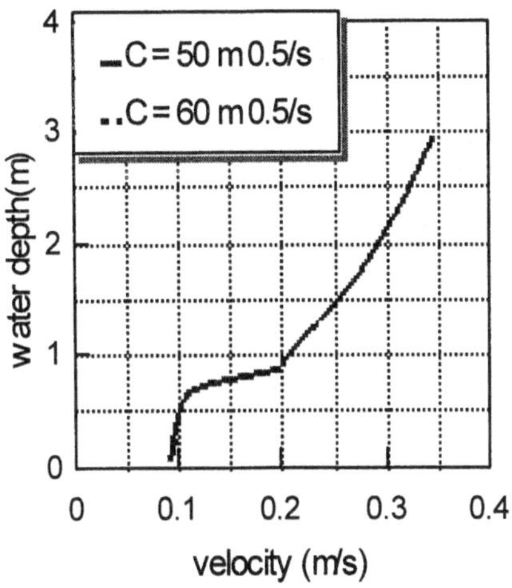

Fig.2.14

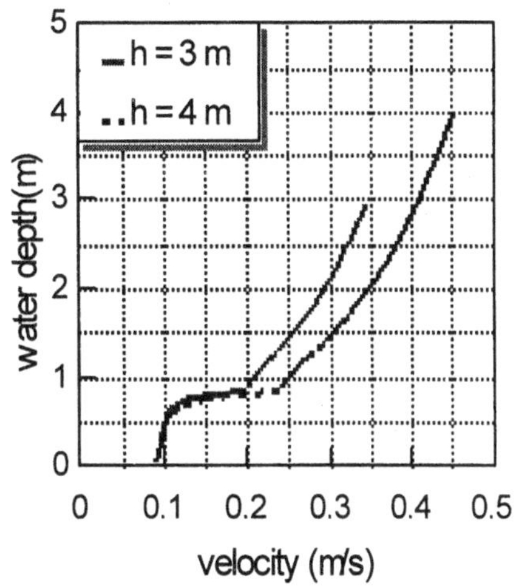

Fig.2.15

The influence of the main hydraulic parameters is presented in Fig.2.14 to Fig.2.16. The Chezy coefficient, whose influence on the velocity profile is presented in Fig.2.14, does not much influence the total resistance; only in the case of a low density of reeds (in this example $m=100$), does it play any significant role at all. The influence of water depth is presented in Fig.2.15. It is seen to act upon the water velocity in the same way as was observed in the case without reeds.

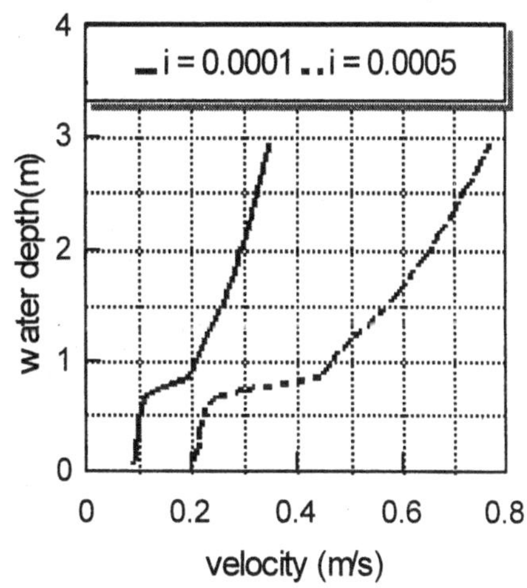

Fig.2.16

The slope of the bottom influences the velocity profile in much the same way with or without the reeds (Fig.2.16). It is clear that it does not impose any influence on the flow resistance in the case of rigid reeds. However, in the case of flexible reeds, higher velocities caused by the steeper bottom slope may cause a greater bending of the reeds. This then results in the reduction of the effective reed height and consequently in the reduction of the additional resistance.

2.6.3 *Parameters connected with the turbulence model*

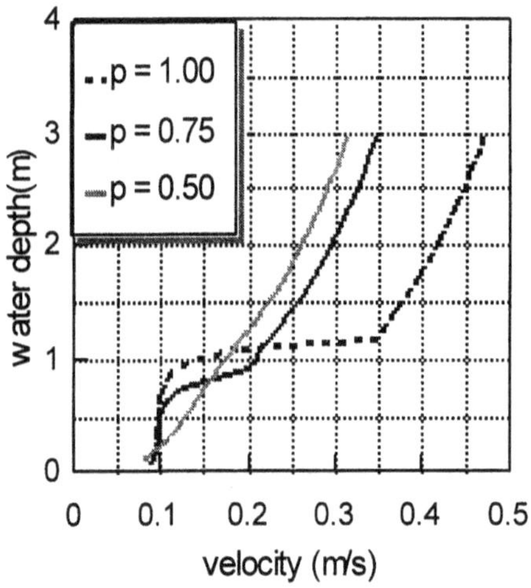

Fig.2.17

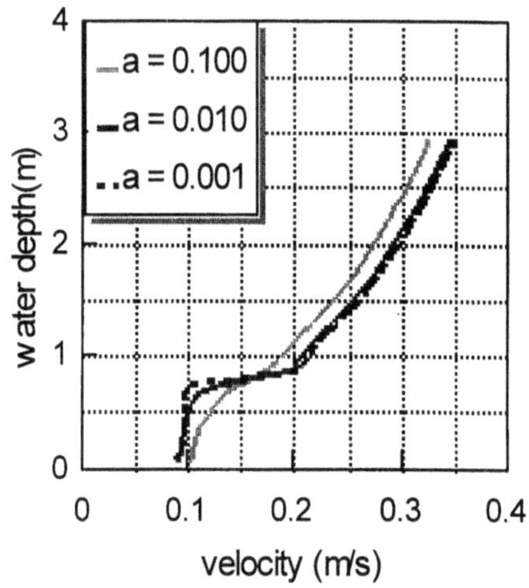

Fig.2.18

The parameter *p,* and in a lesser degree the parameter α connected to the turbulence model, are seen to have a significant influence on the flow resistance: see Fig.2.17 and 2.18. Therefore they should both be calibrated according to the measured data. The significant range for the parameter *α* appears to be between 0.001 and 0.1. It does not appear to influence the results outside of this range.

The parameter *p* introduced in the numerical model plays an important role. As already mentioned, it has a physical origin in the reed geometry and other characteristics. This parameter is a clear example of a parameter which is not independent of other parameters. Further study should be directed to establishing a dimensionally homogeneous functional interrelationship between the parameter *p* and the reed properties.

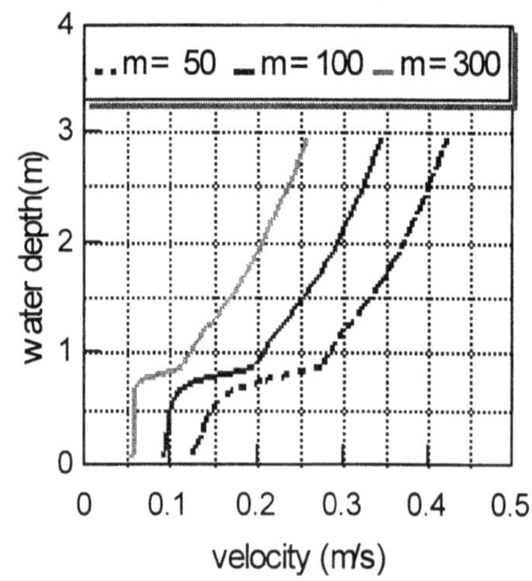

Fig.2.19

2.6.4 *Reed geometry and characteristics*

As already pointed out, the parameter p is obviously dependant on the reed characteristics. Therefore a change of any parameter from this group should as well influence the parameter p. Thus the results obtained by the change of any parameter from this group while the parameter *p* is kept constant do not fully represent the influence of the changed parameter. However, the additional influence exercised through the change in parameter p goes usually in the same direction as does the direct influence of the parameters.

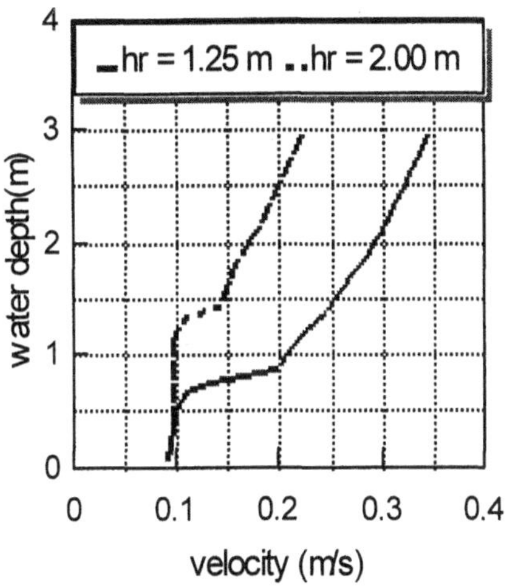

Fig.2.20

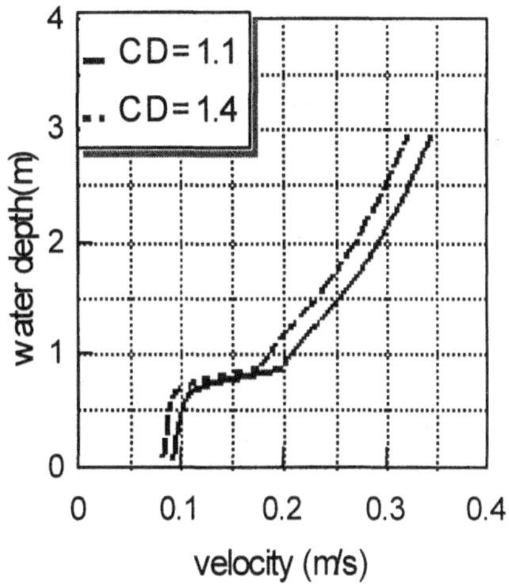

Fig.2.21

The density of the reeds is one of the most important parameters for assessing the additional flow resistances. As can be seen in Fig.2.19, it influences the velocity profile strongly. Another very influential parameter is the height of the reed: as can be seen in Fig.2.20, an increase in reed height decreases the flow velocity quite drastically.

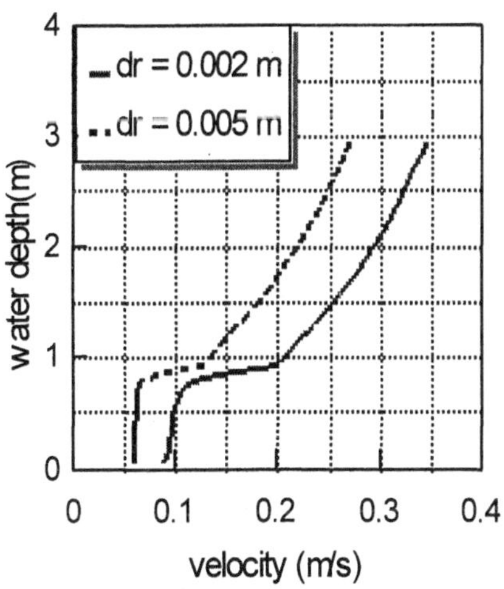

Fig.2.22

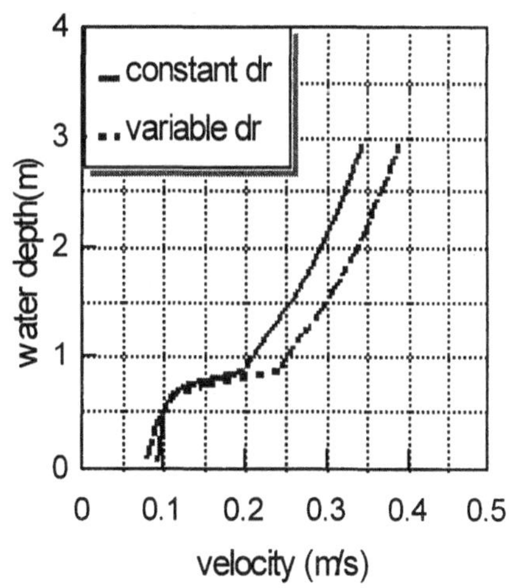

Fig.2.23

In this example, the stiffness of the reed is high, so that it is scarcely bending at all, and in such a case the influence of the reed height is the most strongly pronounced. Fig.2.21 shows the influence of the drag coefficient as estimated from the previous experiments for long rods. Due to the different circumstances obtaining here, this can also be used as a calibration parameter.

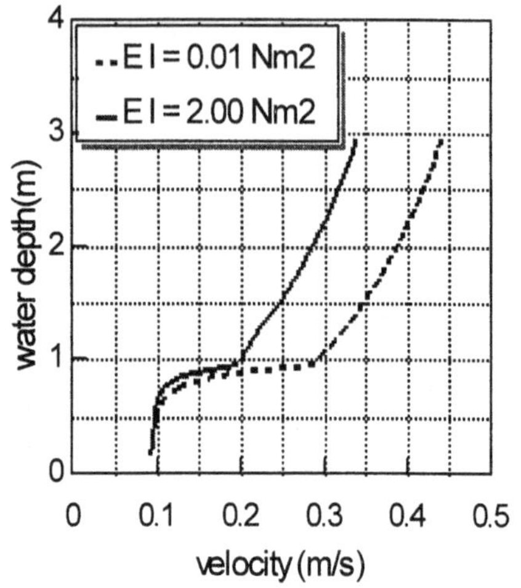

Fig.2.24

The average diameter of the reed plays an important role as well. As it can be seen from Fig.2.22, the smaller diameter causes a lesser resistance. Beside the average diameter, the distribution of the diameter along the reed height is also of importance.

In Fig.2.23, two examples are presented in which the average diameter is the same. In one of them the diameter along the whole reed is constant and equal to the average value. In the other example the diameter at the base of the reed is three time greater than at the top of the reed (while the average diameter is maintained). These results show that a variable diameter causes less resistance to the flow. This result is expected due to the interconnection between flow velocity and the drag force in which the value of the reed diameter is as well included.

In the examples presented in Fig.2.22 and Fig.2.23 the parameter *p* was kept constant. Taking into account that the parameter *p* is not independent of the reed diameter, it can be concluded that the influence of the reed diameter on the additional flow resistance is even greater.

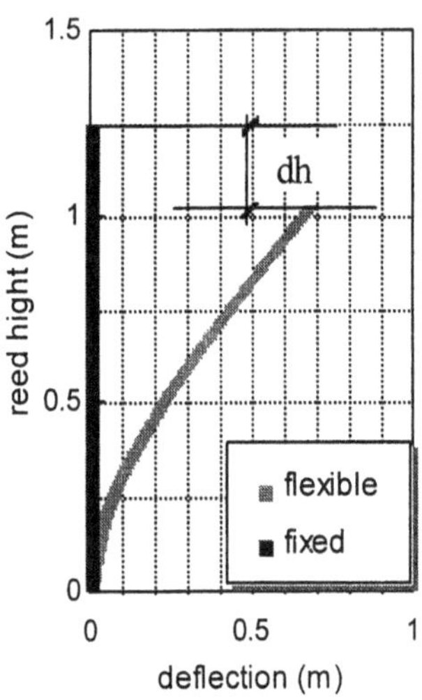

Fig.2.25

The increase of the diameter should cause an increase in *p* and this should decrease the flow velocity even more.

The stiffness of the reed (*EI*) obviously influences the bending capability of the reed. In Fig.2.24 two velocity profiles are presented, one corresponding to the rigid reed and the other to the flexible one, showing, as may be expected, that the flexible reed imposes the lesser additional resistance to the flow.

In Fig.2.25 the deflection of the reed as extracted from the same example is presented. The effective reed height in this example is reduced by the amount *dh*. It is this reduction that largely accounts for the reduction in the additional resistance due to the vegetation.

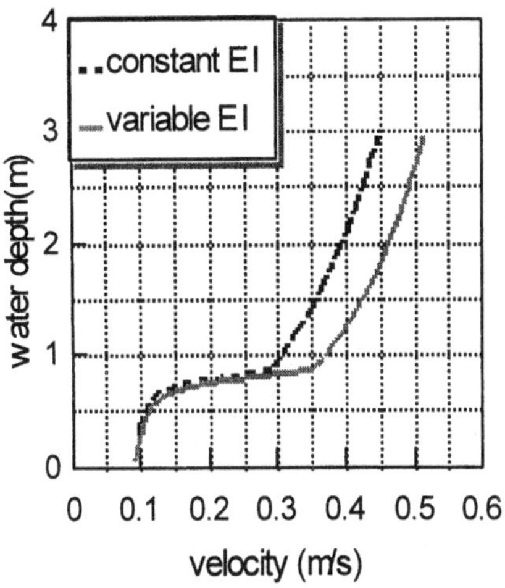

Fig.2.26

As explained earlier, the stiffness of the reed is not expected to be constant along the reed height. Hence the model allows for a linear distribution of the stiffness along the reed. The effects of the stiffness distribution are shown in Fig.2.26 and Fig.2.27. The two examples presented in these figures have the same average stiffness, while one has a constant stiffness and the other a variable one.

From Fig.2.26 it can be seen that reed with a variable stiffness provides the smaller additional resistance. From Fig.2.27, which presents the deflections of the reed in these two cases, it can be seen that the variable stiffness increases the deflection, which in fact means that it reduces the effective height of the reed.

The influence of reed stiffness on the additional flow resistance associated with the parameter p follows the same pattern as the reed diameter. In fact, the reed stiffness is related to the parameter p in such a way that the smaller stiffness causes a reduction of the parameter p and this in turn reduces the additional resistance even more than is immediately evident from these figures.

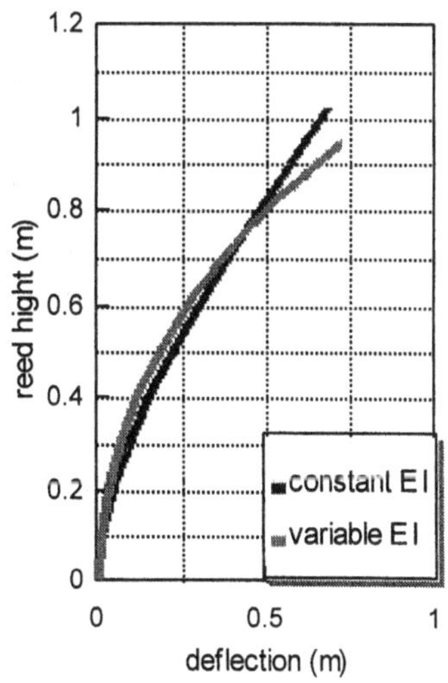

Fig.2.27

2.7 Conclusions and recommendations

The adaptive approach has been shown to be very efficient and transparent and it has provided a basis for the development of this model for the analysis of the influences of flexible vegetation. The *adaptivity* of the scheme was a 'good match' to the *flexibility* of the vegetation. This is an example of how adaptive schemes can solve problems which appear complex in an easy way.

In order to summarise the influences of the various parameters, two cases should first be distinguished: the case of rigid vegetation and the case of flexible vegetation. In the case of rigid vegetation, the most significant parameters are the properties of the height, diameter and density of the vegetation itself.

In the case of flexible vegetation, the role of reed height is taken over by the effective reed height. A significant complexity is introduced by this replacement in the driving mechanism, since the effective reed height is itself a consequence of more parameters. Some of these parameters are again reed characteristics, such as the height, diameter and stiffness of the reed, while others are hydraulic parameters, such as the water depth and the hydraulic gradient. It is obvious that in the case of flexible vegetation the additional flow resistance is a result of more complex interactions again. This model facilitates such further investigations in this field.

Moreover, there is the parameter p defining the portion of the vegetation layer in which the turbulent structure is modelled by an eddy-viscosity approximation, and this appears to play a very important role. While its physical background is still not established, it is suggested to be used as one of the most important of the calibration parameters.

Over and above this, however, the whole model still needs to be verified using field measurements taken in the presence of flexible vegetation. Special attention should then be given to the distribution of the reed stiffness and diameter, as these seem to play an important role in the calculation of the effective reed height. For verification purpose, it would be preferable to measure the effective reed height as well.

This flow adaptive algorithm, once calibrated, can be added, as a module, into an existing two-dimensional nearly-horizontal flow model and in that way form a quasi 3D adaptive model. It would then use flow results obtained by the main program in one time step and return flow resistances to be used in the calculation of flow conditions in the next time step by the main program. However, the complexity of the adaptation transducer can as well be extended if needed.

After calibration, the model can as well be used to produce 'synthetic' data which can be used to produce an emulator or even a formula for the additional flow resistance, and this could replace the model in the cases where its application is considered too time consuming or too complex. Data produced in this way can then be treated by using standard empirical methods with the help of dimensional analysis and by more modern methods, such as neural networks or genetic algorithms (e.g. Minns and Babovic, 1996). The costs of the data produced in this way is negligible in comparison with the costs of the full range of the otherwise necessary field measurements or laboratory investigations. This model is thus a desirable tool for extending the results of physical experiments and field studies to cover a wider variety in the data.

The model presented here provides, in particular, a reliable tool for investigating the additional flow resistance induced by flexible vegetation. It can at the same time fill the gap caused by the scarceness of experimental data in this case, data which are usually difficult to obtain and correspondingly expensive, and so it may help to fill-out even further the ever growing data requirements of modern algorithms as these are applied to water resources management.

CHAPTER 3

Numerical treatment of supercritical flow

3.1 Introduction

One-dimensional unsteady flow in open channels is satisfactorily described by the de Saint Venant equations as long as the hydrostatic pressure distribution assumption holds. Although in the case of discontinuous flows (such as those of hydraulic jumps, dam break fronts and breaking waves) this is not the case in all the points of the domain, the theory of weak solutions (Abbott, 1979, p. 220) shows us that in the integral formulation they can still be used. As these equations cover both sub- and supercritical flow conditions they are also applicable to mixed types of flow. However, there are certain difficulties connected with their solutions. For the numerical modelling of the hydraulic jump, which is the most common case of mixed type flow in open channel hydraulics, there are basically two approaches, namely, the "through" methods based on the theory of weak solutions and usually employing finite difference schemes (i.e. Garcia-Navarro and Saviron, 1992) and the shock fitting methods where shocks are isolated and treated separately (i.e. Rahman and Chaudry, 1995). However, the last faccs practical difficulties connected with tracking the shocks in natural channels (Savic and Holly, 1993).

The through methods based on the method of finite difference can be further divided, following the common division of finite difference methods, into explicit and implicit ones (Abbott and Basco, 1989, pp.213 et seq). The explicit methods, which otherwise suffer severe limitations on the time steps used due to stability conditions, are mainly used for the numerical solution of hydraulic jumps (Chaudry, 1993, pp.203-231). Due to the limited domain on which this physical phenomenon usually occurs, it is still practicle to use a very fine discretisation in the grid, both in time and space. In such cases these explicit methods give satisfactory results.

Due to their stability constrains, however, explicit methods are no longer much used in commercial packages for solution of free-surface flows. Implicit methods are more commonly used in these cases (and then primarily the Preissmann scheme and the Abbott-Ionescu scheme), but these can suffer from another kind of stability problem connected with the modelling of supercritical flows. This has to do with the stability of the algorithmic structure (Abbott and Basco, 1989,pp.151 et seq) in that different algorithmic structures are applicable to sub- and supercritical flows. This makes the solution of mixed type problems quite complex in terms of the control over the code. There are in the recent literature reports of implicit methods capable of solving this problem, such as, for example, the TVD methods (i.e. Garcia-Navarro *et al*, 1994). Still, when compared with the commonly used methods, it is clear that these methods require a considerably heavier computational effort. When applied to the localised problems of hydraulic jump formations they are very efficient, but when applied to the solution of flows in channel networks of considerable size they lead to an unnecessary load.

However, as already mentioned, the schemes of Preissmann and Abbott-Ionescu type encounter another type of problem when dealing with mixed types of flow, and this is connected with the algorithmic structure. In applications, a great simplification is attained when the algorithmic structure can be kept invariant during the whole course of the computation. This is commonly attained by reducing the influence of the convective momentum term when the flow approaches supercritical conditions, so as to maintain subcritical-like characteristic structures everywhere. In this way a flow-adaptive scheme can be developed for the solution of mixed-type flows. In this work, the consequence of using this flow adaptive scheme are investigated and a variant of the momentum equation is identified which reduces the errors introduced through this process to an acceptable minimum.

3.2 Adaptive structure of the algorithm

In order to facilitate the numerical solution of free-surface supercritical flow problems - and indeed, in many cases, to make these practically useful - an adaptation algorithm is introduced which gradually reduces the convective momentum term in the momentum equation as flow conditions approach supercritical (Havnø and Brorsen, 1985; Abbott and Basco, 1989, p.248). This device allows a solution at supercritical flow conditions to be constructed using the same

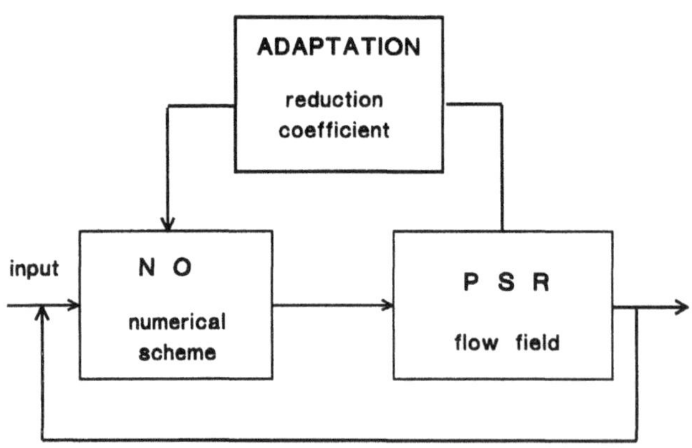

Fig.3.1

algorithm (and boundary conditions) as are used in the case of subcritical flow conditions. By these means, the connectivity properties of any flow network can be kept invariant through a complete computation of mixed subcritical and supercritical flows, thus realising a great simplification at the control level of the code. The corresponding numerical scheme is then a flow-adaptive scheme with quite a simple form of adaptation involved. It is, in its general lines, based on the closed loop adaptive control system and its architecture is represented in Fig.3.1.

The adaptation transducer is just a numerical operator assessing the value of the reduction coefficient for all the points in the solution domain. Being such a simple transducer, and moreover being used at each and every time step, it can be seen as a part of the normal control and not of its adaptivity. However, the fact that it introduces an approximation which facilitates the calculation of the solution but at the same time reduces the accuracy of this solution makes it necessary to distinguish it from the other parameters passed from the PSR to the NO in the control loop.

3.3 Description of the problem

The consequences of excluding the convective momentum term from the momentum equation have to be explored here, especially with reference to the effects of such an exclusion on the boundary conditions. It must in the first place be shown through the study of the characteristic structures how the elimination of the convective momentum term allows the use of boundary condition types that are independent of the flow conditions, and indeed which are of the same type as those required in the subcritical flow case when the full momentum equation is employed.

One-dimensional, free-surface, nearly-horizontal flow is considered and then especially that type of flow that is called supercritical, being that which is characterised by a Froude number that exceeds unity. The Froude number is understood conventionally as a non-dimensional number which describes the ratio between the inertial and the gravitational forces and it accordingly plays an important role in most studies of open channel flow. The NO playing the role of the adaptive transducer uses the Froude number at each discretisation point to assess the values of the reduction coefficient.

Starting from the conservation laws for the mass and momentum and introducing the simplifications of incompressibility of the fluid, a constant density, a hydrostatic pressure distribution and the absence of resistance, we arrive at the de Saint Venant equations in the form:

$$\frac{\partial h}{\partial t} + \frac{\partial (uh)}{\partial x} = 0 \tag{3.1}$$

$$\frac{\partial (uh)}{\partial t} + \frac{\partial (u^2 h)}{\partial x} + gh\frac{\partial h}{\partial x} = 0 \tag{3.2}$$

where:

h	- water depth	g	- gravity acceleration
uh	- discharge per unit width	x	- space coordinate
u	- average flow velocity	t	- time coordinate

The full form of these equations, including the gravity term due to the bottom slope and the friction term, is given in Appendix 3.1.

From these equations is clear that the variables of the PSR are water depth and discharge per unit width. Depending on the numerical scheme used, they are situated either in the same points (non-staggered grid) or in the alternating points (staggered grid) along the domain. The form of the NO is entirely dependant on the finite difference approximation used.

As the analytic integration of this set of equations is not generally possible, they are usually solved numerically. The domain is discretised and, most commonly, partial derivatives are approximated by finite differences. When an implicit method is used, we are then left with a system of quasi-linear algebraic equations. In order to match the number of equations required to the number of unknowns, two more equations have to be added, and these are provided by the boundary conditions (Cunge *et al*, 1980, p.75). Although an implicit finite difference method does not suffer from a time-step limitation for scheme-stability purposes, its algorithmic stability is ensured only if a proper algorithmic structure is used (Abbot and Basco, 1989, p.153). As will shortly be illustrated, this last requirement translates into one of the use of the proper boundary condition relative to the flow conditions and the consequent solution of the system of equations by the corresponding variant of the Gauss elimination method.

The most immediate source of information about the requirements for the boundary conditions to (3.1) and (3.2) is provided by the method of characteristics: the number of characteristics entering the domain in the forward time direction at a point on the boundary gives the number of boundary conditions required at that same point.

The slopes of the characteristic lines defined by equations (3.1) and (3.2) are given by the formula (Abbott, 1966, p.12):

$$\frac{dx}{dt} = u \pm \sqrt{gh} \tag{3.3}$$

From this expression, it is clear that for subcritical flow, where $\sqrt{gh}$ exceeds u, the slopes of the two characteristic lines (C^+, C^-) have opposite signs and therefore enter the domain from two different boundaries, one upstream and the other downstream. For supercritical flow conditions, where the magnitude of u exceeds $\sqrt{gh}$, the slopes of both characteristic lines have the same sign and therefore enter the domain from one side only.

In the case of subcritical flow, one equation is added at each end and thereby a banded matrix of coefficients is formed. For the solution of the resulting equation system, the Gauss elimination method is used in two steps to provide the *double sweep* algorithm (Godunov and Ryabenki, 1964, pp.146-154; Liggett and Cunge, 1975; Abbott, 1979, pp.166-177; Volkov,

1986, pp.155-160; Abbott and Basco, 1989, pp.114-124, 234-248). For supercritical flow, as both equations are added at the upstream end, the whole band in the matrix is situated on one side of the main diagonal. Hence, only the substitution sweep is needed, so as to provide a *single sweep* algorithm.

Problems then naturally arise in the use of implicit schemes when flow in any reach is mixed, so that it changes from subcritical to supercritical or vice versa. Then, in principle at least, a rather complicated adaptive control algorithm with switches should be built in order to allow for the solution of both flow regimes by one, unique algorithm. Although the switching can in its turn be built into the algorithmic structure, the resulting schemes do not appear to have been seriously proposed or reported in the literature (M.B. Abbott, private communication). It has long been observed, however, that supercritical flow conditions can be solved by using the de Saint Venant equations without the convective momentum term with the same algorithm (double sweep) as subcritical flow. Hence the problem has been overcome in practice by introducing a gradual reduction of the convective momentum term as the flow approaches supercritical flow conditions, so that when it arrives at critical flow the equations are employed without any convective momentum term at all. In such a way a simple adaptive control system is built. This control method is implemented and accepted in at least one of the fourth generation modelling systems for one-dimensional, free-surface flow currently available (e.g. Havnø and Brorsen, 1985; Abbott *et al*, 1991).

However, if only because the errors so introduced have been observed to be mostly small and local, little attention has been given to the real performance of the method. Moreover, the question of determining just how the reduction and ultimate absence of the convective momentum term influences the boundary condition requirements and consequently the stability conditions has scarcely been discussed at all. In this chapter at least a part of the answer to this question are to be provided.

3.4 Reduction of the convective momentum term

The method of characteristics is first used to explore the boundary condition requirement in the case when the momentum equation (3.2) is used without the convective momentum term. Here only the simplified form of momentum equation (3.2) is presented while the full form of the de Saint Venant equations is discussed in Appendix 3.1.

In this case, the mass equation (3.1) remains unchanged while the simplified momentum equation (3.2) takes the form:

$$\frac{\partial\,(uh)}{\partial\,t} + gh\frac{\partial\,h}{\partial\,x} = 0 \tag{3.4}$$

In matrix form, the two equations, (3.1) and (3.4), can be written as:

$$\frac{\partial f}{\partial t} + A\,\frac{\partial f}{\partial x} = 0 \tag{3.5}$$

where:

$$f = \begin{bmatrix} h \\ uh \end{bmatrix} \quad , \quad A = \begin{bmatrix} 0 & 1 \\ gh & 0 \end{bmatrix}$$

The eigenvalues of the matrix A give the slopes of the characteristic lines :

$$\frac{dx}{dt} = \pm\sqrt{gh} \tag{3.6}$$

From this expression, it is seen that slopes of the characteristic lines do not depend at all on the Froude number. Moreover, the two characteristic lines always have opposite signs, which corresponds to the use of a one-point boundary condition at each end. Thus, the double sweep algorithm is the proper algorithmic structure for all flow conditions in this approximation: models with a built-in gradual reduction of the convective momentum term and one point boundary data at each boundary are here stable even in the case of supercritical flow.

If, instead of the conservation of momentum, the conservation of the momentum per unit volume is considered, using equation (3.1) and equation (3.2) and assuming that all functions are differentiable, we arrive at the Bernoulli equation:

$$\frac{\partial u}{\partial t} + u\,\frac{\partial u}{\partial x} + g\,\frac{\partial h}{\partial x} = 0 \tag{3.7}$$

The full form of equation (3.7) given as well in Appendix 3.1. In combination with the mass equation (3.1), the Eulerian matrix form can be written as:

$$\frac{\partial f}{\partial t} + A\,\frac{\partial f}{\partial x} = 0 \tag{3.8}$$

where:

$$f = \begin{bmatrix} h \\ u \end{bmatrix} , \quad A = \begin{bmatrix} u & h \\ g & u \end{bmatrix}$$

When the full equations (3.8) are used, the slopes of the characteristic lines naturally have the same slopes as in the case when the full momentum equation is exploited.

If, however, the convective term is eliminated from the Bernoulli equation component of the matrix A in (3.8), we obtain:

$$A = \begin{bmatrix} u & h \\ g & 0 \end{bmatrix}$$

It is important to observe that this matrix could not be obtained from the continuity equation (3.1) and the momentum equation with reduced convective momentum term (3.4), so through this transition a novel form is obtained.

The eigenvalues which define the slopes of characteristics are now given by:

$$\frac{dx}{dt} = \frac{u}{2} \pm \sqrt{\frac{u^2}{4} + gh} \qquad (3.9)$$

(For details see Appendix 3.1). From this expression for the slopes of the characteristic lines, it is apparent that these are not independent of the flow velocity. The slope of the C^- characteristic remains negative no matter what the relation between u and $\sqrt{gh}$; but its magnitude decreases with the increase of the Froude number. This is a property which can be used to influence results in a felicitous direction. When flow is supercritical the downstream boundary condition is, strictly speaking, physically unrealistic, and therefore completely artificial, but this property has the consequence that the influence of this departure from reality should not be spread very widely through the domain. By virtue of the very steepness of the C^- characteristic (with flow taken in the positive direction), its influence is limited to a very narrow region.

If the C^+ characteristic is considered (3.9) it is obvious that the relative error in its slope compared with the slope of the C^+ characteristic of the full equations (3.3) decreases with the increase in the Froude number.

In Fig.3.2 the 'relative slope', defined as the ratio between the slope of the characteristic line and that of $\sqrt{g\,h}$, is presented as a function of the Froude number for all cases considered here. Let us first consider the case where the complete equations are used (presented by full lines). For supercritical flow, when the Froude number exceeds unity, both characteristics are seen to have a positive slope, which leads to the appropriateness of two-point boundary condition data at the upstream end.

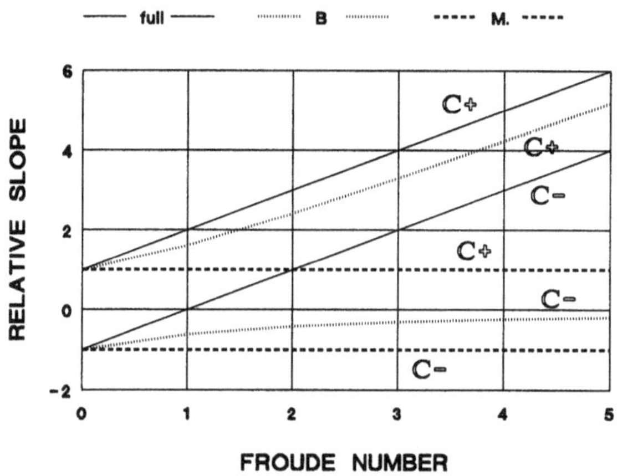

Fig.3.2 Relative slopes of the characteristic lines as a function of the Froude number

B=Bernoulli equation without the term $u\dfrac{\partial u}{\partial x}$

M= momentum equation without the term $\dfrac{\partial}{\partial x}\left(\dfrac{u^2}{h}\right)$

The slopes of the C^- characteristics in both of the other cases considered, where the equations without the convective terms are used, remain negative in the supercritical flow conditions, which makes possible the use of one-point boundary condition data at both ends of the domain. However, a distinction between these two C^- characteristics is possible. The one arising from the use of the momentum equation (dashed line) has a constant slope, while the other, derived from the Bernoulli equation (dotted line), tend asymptotically towards zero with increasing Froude number, what is to say that this slope gets increasingly steeper in the physical plane. As already mentioned, the use of the steeper slope introduces smaller errors from the use of the, strictly-physically-incorrect, downstream boundary condition.

The slopes of the C^+ characteristics for the cases without the convective term also demonstrate large differences. In order to explore these differences in more detail, a 'celerity coefficient' is defined as the ratio between the celerity of the set of equations actually employed and the celerity of the full equations. In Fig.3.3, the celerity coefficient is presented as a function of the Froude number.

The celerity coefficient of the momentum equation without the convective term (dashed line) shows a very rapid drop with the increase in Froude number, and this leads to the conclusion that the celerity of the solution obtained by that method differs much too drastically from the celerity of the solution obtained with the full equations. On the other hand, the celerity coefficient of the Bernoulli equation without the convective term (solid line) reaches its minimum of approximately 0.8 at a Froude number of around 1.6, and this is followed by a gradual improvement in representative capability. Therefore, no very significant error in celerity is to be expected in this case.

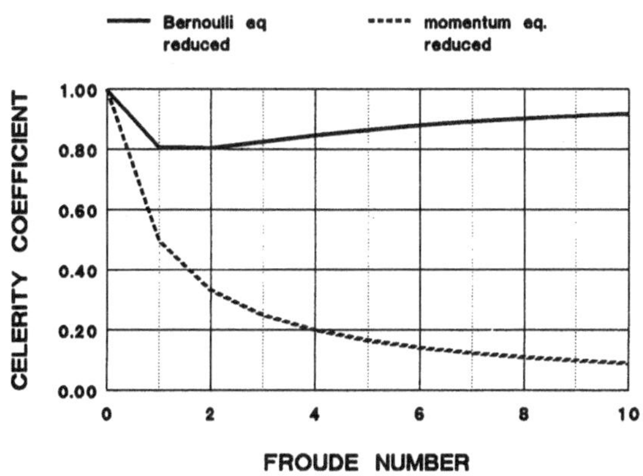

Fig.3.3 Celerity coefficient as a function of the Froude number

However, the celerity of the solution was not the only feature to be considered, as it will be shown later. The celerity ratio was used here as an example of the schemes behaviour as it demonstrates this particularly clearly.

3.5 Which term is to be reduced?

All the differences shown in the previous section between results obtained with the momentum and Bernoulli equations, both without convective momentum terms, lead to the conclusion that the reduction of the convective momentum term from the momentum equation is excessively severe. A reduction of the convective term in the Bernoulli equation appears to suffice to allow the use of the same algorithmic structure for the solution of the subcritical and the supercritical flow. Moreover, it promises to provide a much closer solution to the solution of the full equations. As the momentum equation is the more commonly used in the modelling of natural rivers, the notion naturally arises of splitting up the convective momentum term and reducing only a part of it. One possible expansion of the term is then:

$$\frac{\partial\,(u^2h)}{\partial\,x} = hu\frac{\partial\,u}{\partial\,x} + u\frac{\partial\,(uh)}{\partial x} \tag{3.10}$$

If only the first term of this expansion (which anyhow resembles the convective term from the Bernoulli equation) is reduced, expression (3.5) takes the form:

$$\frac{\partial f}{\partial\,t} + A\,\frac{\partial f}{\partial\,x} = 0 \tag{3.11}$$

where

$$f = \begin{bmatrix} h \\ uh \end{bmatrix}, \qquad A = \begin{bmatrix} 0 & 1 \\ gh & u \end{bmatrix}$$

(For more details see Appendix 3.1.)

The eigenvalues of the matrix A, which define the slopes of the characteristics, are then given by:

$$\frac{dx}{dt} = \frac{u}{2} \pm \sqrt{\frac{u^2}{4} + gh} \tag{3.12}$$

which is the same expression as obtained in the case that the Bernoulli equation is used. Accordingly, all the positive features of the Bernoulli equation formulation can be exploited in this case as well.

3.6 Stability analysis

The stability analysis for a numerical solution obtained by any finite-difference method is usually done by means of a linearised analysis in Fourier series expansions. In such an analysis, the amplifications of the wave components of the unknowns within one time step are examined. The analysis is highly localised and does not consider the application of the boundary conditions. Thus, although stability proved in this way is essential for the performance of any scheme, it is by no means sufficient, and this is especially so in the case of implicit schemes. Usually it is accompanied by further stability conditions, commonly taking account of the required boundary conditions, such as may be obtained from the method of characteristics. In this case a distinction is sometimes made between the stability of the *scheme* and the stability of any particular *algorithm* (or ordering of operations) used to solve the scheme (see Abbott, 1979, pp.178-182; Abbott and Basco, 1989, pp.78-86, 251-259). Examples of both of these kind of analyses are presented in Appendices 3.5 and 3.6. It should further be observed that stability conditions will depend upon the form in which the conservation laws are written: a proof of stability for a scheme based on (3.4) is no guarantee of stability for 'the same' scheme based on (3.7), and vice versa.

3.7 Models

Numerical solutions of the de Saint Venant equations are most commonly obtained by using finite difference methods. In practice, two difference schemes are mainly used for free-surface flow simulations, namely, the Preissmann scheme and the Abbott-Ionescu scheme. For details on finite difference approximations see Appendices 3.2 and 3.3 while the details of the solution algorithms are given in Appendix 3.4. Models based on both of these schemes were developed in order to test the influences of the different reductions of the convective momentum terms on the results.

The model which uses the Preissmann scheme for the finite-difference approximations was based on the Eulerian form of the equations. The non-staggered grid which it uses makes possible the direct approximations of partial derivatives of both unknowns (water depth and velocity or discharge) with respect to both arguments (time and space).

For the staggered grid of the Abbott-Ionescu scheme, the finite difference approximation of a partial derivative with respect to space of only one unknown within one component of the scheme is possible. Accordingly, equations of the form of (3.5) have been rewritten in their algorithmic form:

$$A^{-1}\frac{\partial f}{\partial t} + \frac{\partial f}{\partial x} = 0 \ , \qquad |A| \neq 0 \ ,$$

where, for the momentum form, for example:

$$f = \begin{bmatrix} h \\ uh \end{bmatrix} \quad , \quad A^{-1} = \frac{1}{u^2-gh}\begin{bmatrix} 2u & -1 \\ u^2-gh & 0 \end{bmatrix}$$

The algorithmic forms are derived in a similar way for all the other cases of the reduced or partially-reduced equations, as well as for the cases where the Bernoulli equation is used (see Appendix 3.1). Moreover, two more terms were introduced, in both momentum and Bernoulli equation, to cover the bottom slope and the friction effects of the channel.

For each of the above schemes, four models were developed for supercritical flow, denoted as follows :

SS - based on the complete equations and consequently with a two-point boundary condition at the upstream end.

B - based on the mass equation and the reduced Bernoulli equation and so with a one point boundary condition at each end

M1 - based on the mass equation and the fully-reduced momentum equation and so with a one point boundary condition at each end

M2 - based on the mass equation and partly-reduced momentum equation and consequently again with a one point boundary condition at each end

The role of the models (SS) based on the full equations, and consequently solved using a single-sweep algorithmic structure from two point boundary data at the upstream boundary, provides reference results. The accuracy of these results is not treated in this work, this having been discussed extensively elsewhere (see, for example, Abbott, 1979, pp.180, 185-188; Abbott *et al*, 1981; Cunge *et al,* 1980, p.89; Abbott and Basco, 1989, pp.257-259).

3.8 Results

In all the models based on the equations with a reduced convective momentum term and consequently with one-point boundary data at each end (B, M1 and M2), oscillations were observed. These started gradually, but in time built up to cause instabilities. The origin of this noise was traced to the form of the friction term and accordingly the cases with higher Froude numbers were influenced more. In order to avoid this effect, a dissipative interface was built into each of the models through a simple averaging of the parameters (flow velocity, area of the cross-section, etc.). This can as well be recognized as a form of an adaptive algorithm as was already mentioned in Chapter 1. For models based on the Bernoulli equation, this proved to be sufficient, while for the models based on the momentum equation a forward-in-time centring of the schemes was also employed. This appears to be in agreement with current practice (Abbott and Basco, 1989, p.248; DHI, 1987, p.C-2-6). All results presented from the

SS and B models were obtained using fully-centred schemes, while the M1 and M2 models employed schemes centred completely forward in time, and so with $\theta=1$.

The example for which the results are presented is an unsteady flow in a 200m long, rectangular, 10m wide channel with a uniform bottom slope of 0.01 and a roughness defined by a Chezy coefficient of $50m^{1/2}s^{-1}$. The initial condition is a uniform supercritical flow with 0.5m depth and a Froude number of 1.6. The upstream boundary condition is a water depth that is set as a function of time, such that in 60s this depth increases linearly from an initial 0.5m to 1.0m, and then, within the next 60s returns to 0.5m, and after that stays constant. The downstream boundary condition is a rating curve based on uniform flow conditions as defined by the bottom slope and the Chezy coefficient. The computational parameters were $\Delta t=10$s and $\Delta x=10$m for the Abbott-Ionescu scheme and $\Delta t=10$s and $\Delta x=20$m for the Preissmann scheme, which give in both cases a Courant number of approximately 6.

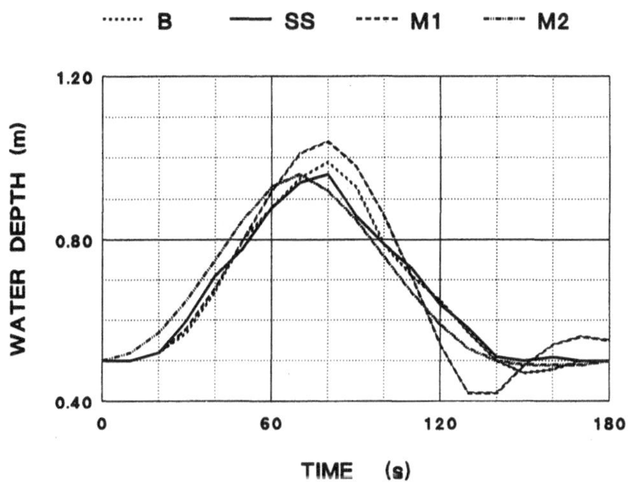

Fig.3.4 Water depth as a function of time at point $x=100$m, Abbott - Ionescu scheme

As the results obtained by the two schemes used here do not differ substantially, the same conclusions hold for both schemes.

In the results obtained by all the models employing the reduction of the convective momentum term, the water depths are overestimated (Fig.3.4) if compared with the reference results of the SS model. The results of the B model are the closest to the reference ones, specially if one takes into account that by off-centring the scheme in space they can be brought even closer together. At the same time, it is clear that results for the water depths given by the M1 model are the least accurate.

Results for discharges are presented in Fig.3.5. Again, those obtained using the B model are the closest to the results of the SS model. The values obtained using models M1 and M2 are too low, mainly due to the numerical diffusion introduced by the forward-in-time centring of the finite difference schemes. The phase error, which can be seen clearly in all three cases (B, M1, M2), was expected, and especially in the case of M1 model, which shows the greatest error of this kind. This error is not primarily a consequence of the numerical solution, but is due to the different celerities (slopes of the characteristics) of the differential equations upon which each of the models is based. As introduced above, it is caused by the partial or full reduction of the convective term. For a Froude number between 1.6 and 2.0, as in this example, the results of models B and M2 show the greatest phase error, which then decreases with increasing Froude number. However, the celerity error of the model M1 increases monotonically with increasing Froude number.

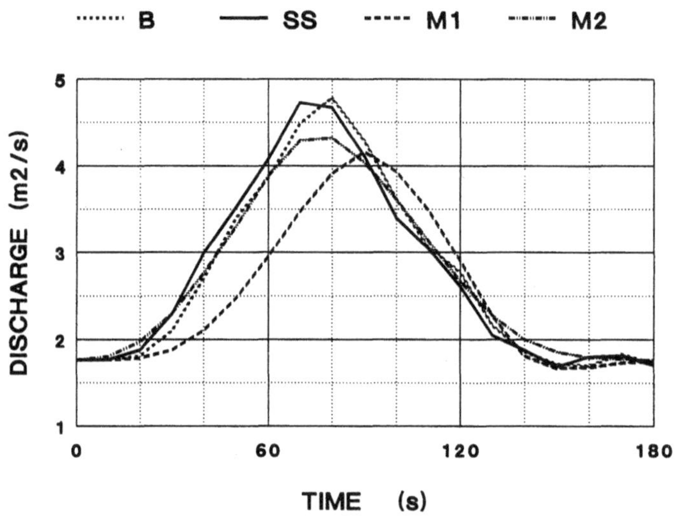

Fig.3.5 Discharge as a function of time at point $x = 100$m, Preissmann scheme

The results for velocities, as shown in Fig.3.6, demonstrate how in the cases of models B and M2 these are underestimated, so that there is again a phase error. On the other hand, the results of the M1 model are quite highly unrealistic. In fact, the velocities in this case were calculated as the ratio between the discharges and the areas of the cross-sections: as the discharges demonstrated much larger phase errors than the water depths, their ratio has little physical sense, while the velocities so calculated are completely out of the reasonable range.

The overall underestimation of velocities in the case of a partly or completely reduced convective term has also a return effect on the reduction itself. The reduction coefficient is defined as a function of the Froude number, which is obtained using the calculated velocities. However, the final effect is positive, while underestimated velocities give smaller Froude numbers, which then cause smaller reductions of the convective term and consequently more accurate calculations of velocities.

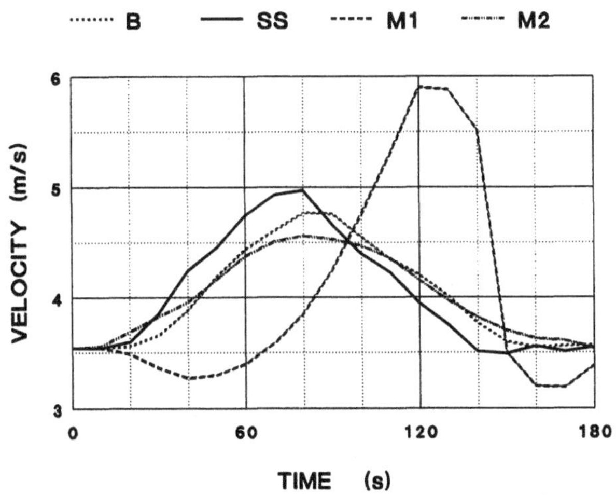

Fig.3.6 Velocity as a function of time at point $x=100$m, Abbott-Ionescu scheme

In the current practice of free-surface modelling, the most commonly used set of equations is that composed of mass and momentum equations. The reduction of the convective momentum term leads to model M1. If now, through Fig.3.4 - Fig.3.6, the results obtained in this case are compared with the ones obtained from the SS model, taken as a reference, it can be seen that these differ considerably. The results for water depths are the most overestimated, discharges are the most reduced and delayed in time, while the velocities become correspondingly even more unreliable.

3.9 Conclusions

In this chapter several simple flow adaptive algorithms were investigated. All these adaptive algorithms were developed with the aim of modelling supercritical and mixed flow conditions. They were based on different formulations of conservation laws and different finite difference approximations; moreover, they have used different forms of adaptation. The main benefit from their adaptivity is that they provide an approximate solution in the case where other algorithms would fail or become excessively expensive in terms of machine time and code complexity. This approximate solution is of course inferior to the numerical solution devoted entirely to the problem of supercritical flow, but from the point of view of overall solution efficiency, the use of such methods appears to be justified. However, if this should be required, once the adaptive algorithm has identified the reaches where supercritical flow occurs, the more detailed models can be applied on these limited sub-domains.

When the convective momentum term is excluded from the momentum equation, the slopes of the characteristic lines become independent of the flow conditions and depend only on the flow depth so that they inevitably preserve the integrity of the subcritically-posed boundary conditions. However, with an increase in Froude number, the differences between the slopes of these characteristics and those of the full equations increase and this naturally has its consequences for the celerity of the numerical solution.

Despite these obvious limitations, however this approach has proved itself to be stable, simple and flexible. It has been shown that, even for supercritical flow conditions, the resulting use of one-point boundary data at each boundary makes numerical schemes solvable using the same algorithm as is used in subcritical flow (the double-sweep algorithm). The method is unconditionally stable in the primitive form, and, so long as one remains aware of the range of its accuracy, it can be very useful. In general, all results were shown to suffer from much the same time delay, while water depths were too high and discharges and velocities were too low. A numerical diffusion can be employed to improve the results for the water depths, but this further reduces the reliability of results for discharges and velocities. An increase in steepness of the channel bottom, however, did improve the accuracy of the models.

If, instead of the momentum equation, the Bernoulli equation is used, and its convective term reduced, the expressions for the characteristics, that in turn govern the algorithmic structure, continue to differ. Their direction is still independent of the flow conditions, but they match with the original expressions more satisfactorily. Hence, the induced error in the celerity is much smaller and it decreases monotonically with the increase in flow velocity. Moreover, the slope of the characteristic proceeding in the opposite direction to the flow is much steeper, so that it limits the influence over time of the artificially-posed downstream boundary condition to a narrower region.

In general, when the momentum equation is used with the fully-reduced convective momentum term (M1), the results are by far the worst in comparison with the other cases, as is in agreement with the theory. Therefore, the full reduction of the convective momentum term should be avoided. Since, however, changes resulting from excluding the convective term in the Bernoulli equation suffice to ensure the stability and influence less the accuracy of the solution, in the momentum equation, which is mainly used in practical applications, a reduction of only a part of the convective term, which corresponds to the convective term of the Bernoulli equation, is proposed. This appears to be sufficient for the purpose of maintaining stability while it influences the solution accuracy to a minimal degree.

CHAPTER 4

A generalised solution algorithm for flow in channel networks

4.1 Introduction

When implicit finite difference numerical schemes are applied to solve numerically one-dimensional flow equations, three algorithmic structures can be identified as functions of the flow characteristics: a single left-to-right sweep, a double sweep and a single right-to-left sweep, corresponding to a left-to-right supercritical flow, a subcritical flow, and a right-to-left supercritical flow (Abbott and Basco, 1989, pp.208-210). In practice however it has just been shown that there are computational-organisational advantages in approximating the supercritical flows by flows that necessitate the use of only one algorithmic structure, namely that which is properly appropriate to subcritical flow (see also Kutija, 1993).

The application of these three different structures within the same model leads to a series of complications as soon as networks of individual channels are considered. It is then customary to divide such networks into two classes, the first being of the tree-like, branched or dendritic kind and the second being of the looped or multiply-connected kind, as schematised in Fig.4.1. In the first of these, a), the extension of the single-channel algorithm is effected relatively simply by a suitable ordering of the initiating sweeps of the algorithm, as again schematised in Fig.4.1 and as explicated elsewhere (Cunge *et al*, 1980, pp.109-113).

In the second kind, as schematised in Fig.4.1b, however, no such simple solution is possible and other approaches must be taken. The most widely used of those consists of eliminating the influence of the points intermediate between the nodes of the looped network so as to reduce the problem effectively to one of inverting a matrix the rank of which is simply the number of nodes (Cunge *et al*, 1980, pp.113-121).

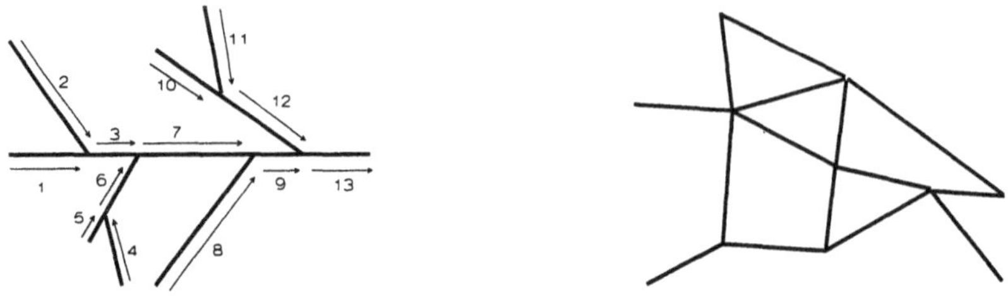

a) branched channel network b) looped channel network

Fig.4.1

This process is again simplified if double-sweep algorithms are used uniformly. Then, in view of the availability of a number of 'fast solvers' for the linear equations of the resulting 'well-conditioned' matrices, these standard tools can be conveniently used to solve for the nodes, and thence for the points between these nodes. However, most of these 'solvers' use iterative methods which are not only usually slower than direct, noniterative procedures for complete networks, but the use of which also tends to weaken the user's insight into the links between the algorithmic structures of the solutions and the physical information-transmitting properties of the physical system. Moreover, the computational time used commonly increases more than linearly with the number of nodes in the case of looped networks (e.g. Osiadacz, 1987, p.89). This last feature can be a major handicap when solving for larger urban drainage systems for real-time control purposes, since such a system typically consist some tens and occasionally some hundreds of loops, and each control setting necessitates at least one computation for every effective, physical control structure or other element (pump, syphon, weir, etc.) in the system. Although essentially logical methods may be used to break up the system into a set of more manageable sub-systems and a multi-processing capability may be employed to compute each such sub-system and the influence of each physical control structure or pump in parallel, a useful benefit still accrues from using faster direct methods.

The general network algorithm presented here provides a fully-automated instantiation and control procedure for combined dendritic and looped networks that is physically transparent and which provides solution speeds that are little influenced by the network complexity. Although explicated here only for the case of flows approximated using double-sweep algorithmic structures, it seems possible to extend these methods to other, including mixed, algorithmic structures should that later be deemed necessary. In the same vein, it is supposed for the moment that the topology of the network is made invariant by the usual procedure of introducing Preissmann and Abbott slots (Abbott and Basco, 1989, p.248). The use of the device of Preissmann slot also allows the methods described here to be applied directly to fully pressurised flows and thus to water distribution and associated water hammer problems (Vanecek *et al,* 1994).

4.2 Adaptive control structure

The control mechanism of the generalised network algorithm presented here is provided by a graph-theoretical approach to the network topology. It introduces two novel aspects. The first is the definition of a control sequence which takes into account the physical information-transferring properties of the network through a graph-theoretical formulation. The second is the introduction of a global elimination algorithm for the solution of the looped network which uses the predefined control sequence and employs automatised symbolic computation for the solution of the large sparse system of linear equations that is produced.

The control mechanism applied in this algorithm plays a very important role while it provides the order of calculation. Thus, the results provided by the adaptation transducer forms an essential part of the numerical operator. Without the lists defining the sequence of calculations, the numerical operator could not perform any useful calculation at all.

The control mechanism provides, firstly, the distinction between the two different types, and corresponding parts of the network from the point of view of calculation so that the most efficient algorithm can be applied on each of these parts, both dendritic and looped. Secondly, it provides an order of calculation for each or these algorithms that corresponds to the topology of the network. The details of this control mechanism are discussed later in this work.

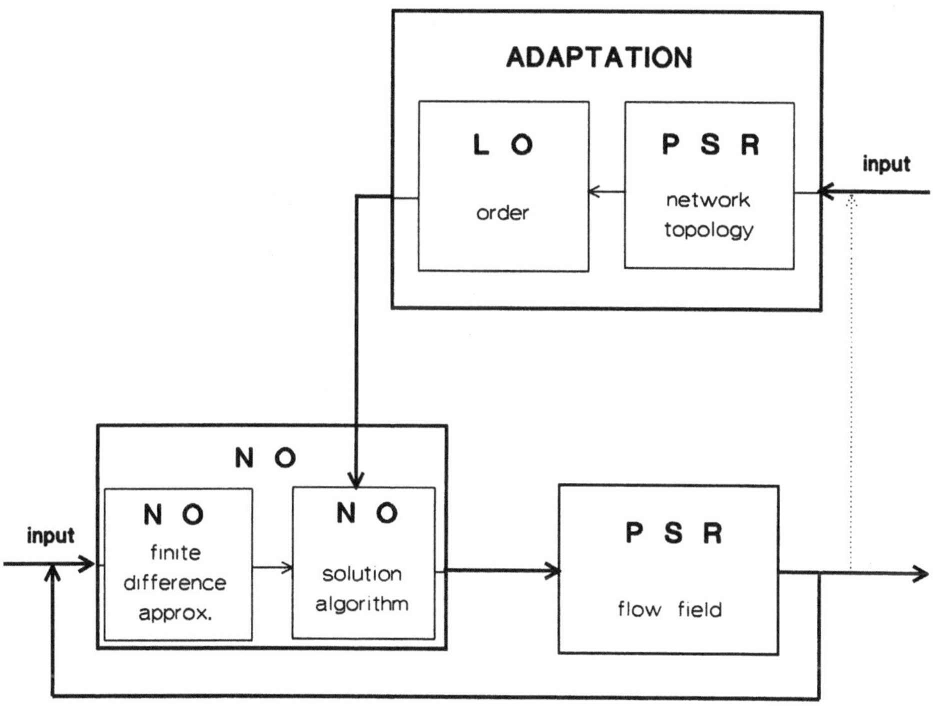

Fig.4.2

The adaptive nature of this control mechanism is of a type where the process of adaptation of a generally given method is prepared at the beginning of the operation according to the specific features of the problem. Thus, in effect, the definition of the control sequence *is* the adaptation process. In this way the control mechanism is defined on the basis of network topology before the time dependant calculation. If however, the network topology changes during the simulation (as a consequence of the drying-out of a channel or the opening or closing of a valve or a gate) the corresponding adaptation of the control mechanism can be introduced. The adaptation transducer is composed of a PSR-network topology analyser and a logical operator (LO) which defines the order which is followed in the solution algorithm.

The scheme of this adaptive control system is represented in Fig.4.2. From the scheme it becomes obvious that this adaptive control system can be characterised as one of partly open loop adaptation. This is the case if the results of flow simulation (states of PSR hydrodynamics) are not seen as having an influence on the network topology (that is the PSR network topology). If however, the network topology depends on the flow results, then it becomes a closed loop control system.

The numerical operator (NO) itself is considered as a transducer which is itself composed of two essentially different transducers, the one being concerned with the numerical discretisation and controlled by the results obtained at previous time step (that is states of the PSR hydrodynamics) and the other being the solution algorithm controlled by the lists originating from the adaptation transducer.

As the proposed algorithm follows the main line of thinking of computational hydraulics, in that the order of computation and associated recursion is based on the physics of the channel flow, and corresponding information flows, it is expected to be more efficient than the iterative algorithms that are still the most commonly employed. Although algorithms providing direct-solutions have been constructed and applied in some cases since the 1970s (e.g. DHI, 1992), and most recently by Verwey (1994; see also Vanecek *et al*, 1994), no general theory for such procedures appears to have been previously elaborated.

4.3 Description of the problem

4.3.1 *Elements of graph theory*

Only the barest elements of graph theory are introduced here by way of background (Chen, 1990, pp.1-10). In general we shall understand by a graph G an ordered pair of disjoint sets (V,E) such that E is a subset of the set of unordered pairs of V, where:

V is the set of vertices

$$V = \{ 1 , 2 , 3 , 4 \}$$

E is the set of edges

$$E = \{ \{1,2\} , \{1,4\} , \{2,4\} , \{2,3\} \}$$

$G(V,E)$ is a graph

$$G = \{\{1,2,3,4\} , \{\{1,2\},\{1,4\},\{2,4\},\{2,3\}\}\}$$

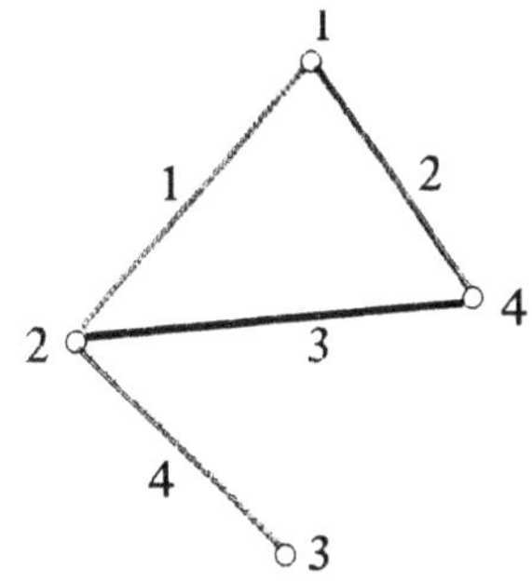

Fig.4.3

Any graph can as well be represented by a connected set of points, lines and numerals (see Fig.4.3) but the usual way of presenting a graph is by its *incidence matrix* (Chen, 1990, p.53). The incidence matrix of a graph is a matrix $B = B(n \times m)$, with n being the number of vertices and m being the number of edges, defined as:

$$B = B(G) = \{b_{ij}\}, \quad b_{ij} = \begin{cases} 1 \text{ if } v_i \text{ is a vertex of an edge } e_j \\ 0 \text{ if } v_i \text{ is not a vertex of an edge } e_j \end{cases}$$

Thus the incidence matrix of the graph of Fig.4.3 is:

$$\begin{array}{c|cccc} i \backslash j & 1 & 2 & 3 & 4 \\ \hline 1 & 1 & 1 & 0 & 0 \\ 2 & 1 & 0 & 1 & 1 \\ 3 & 0 & 0 & 0 & 1 \\ 4 & 0 & 1 & 1 & 0 \end{array}$$

Each column in the incidence matrix represents an edge, so that, as each edge has two ends, the sum of all the entries in any one column is equal to two. This property is usually used for checking the input data. Each row represents a vertex and the sum of all the entries in one row is called the *degree* of a vertex. The degree is the number of edges incident to that vertex. A vertex with a degree equal to one, like vertex number 3 in this example, is an end vertex. A graph which can be drawn in the plane in such a way that no two edges intersect is called a *planar graph* (Chen, 1990, pp.29-36).

In the following expositions, channel networks will be treated as planar graphs. This abstraction will not cause difficulties because any channel network resembles a graph by virtue of its one-dimensionality. In hydraulics the vertices are commonly called 'nodes' or 'junctions', while the edges constitute the 'links', or 'channels', or 'pipes' or 'conduits', depending upon the area of application concerned. For the sake of generality, we shall keep to the relatively neutral terms of 'vertices' and 'edges'; these can then be easily transcribed into the particular terms that are used in any particular hydraulic context. It remains sufficient that the network topology can always be expressed in terms of the incidence matrix, as in this manner all the connectivity relations between the elements of the graph are given in an unique and recognisable way.

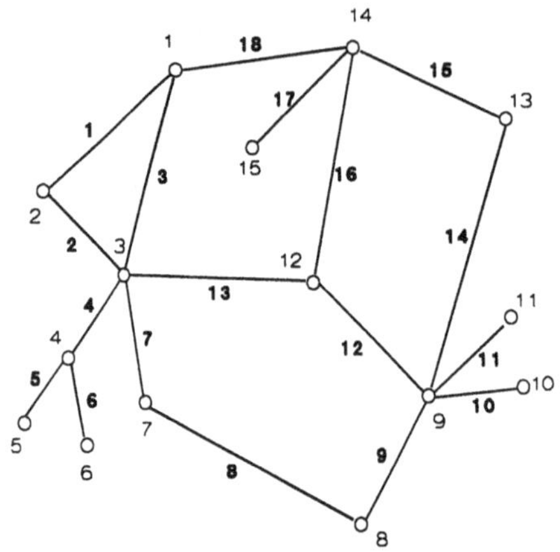

Fig.4.4

$$
\begin{vmatrix}
1&0&1&0&0&0&0&0&0&0&0&0&0&0&0&0&0&1\\
1&1&0&0&0&0&0&0&0&0&0&0&0&0&0&0&0&0\\
0&1&1&1&0&0&1&0&0&0&0&0&1&0&0&0&0&0\\
0&0&0&1&1&1&0&0&0&0&0&0&0&0&0&0&0&0\\
0&0&0&0&1&0&0&0&0&0&0&0&0&0&0&0&0&0\\
0&0&0&0&0&1&0&0&0&0&0&0&0&0&0&0&0&0\\
0&0&0&0&0&0&1&1&0&0&0&0&0&0&0&0&0&0\\
0&0&0&0&0&0&0&1&1&0&0&0&0&0&0&0&0&0\\
0&0&0&0&0&0&0&0&1&1&1&1&0&1&0&0&0&0\\
0&0&0&0&0&0&0&0&0&1&0&0&0&0&0&0&0&0\\
0&0&0&0&0&0&0&0&0&0&1&0&0&0&0&0&0&0\\
0&0&0&0&0&0&0&0&0&0&0&1&1&0&0&1&0&0\\
0&0&0&0&0&0&0&0&0&0&0&0&0&1&1&0&0&0\\
0&0&0&0&0&0&0&0&0&0&0&0&0&0&1&1&1&1\\
0&0&0&0&0&0&0&0&0&0&0&0&0&0&0&0&1&0
\end{vmatrix}
$$

In the same vein, the complexity of the graph can be measured with different parameters, which however can be used uniformly in all areas of application in hydraulics. The *first Betti number* $(m\text{-}n+1)$ defines the *number of loops* in a graph (Abbott, 1979, p.216). Another measure is the ratio between the number of edges and the number of vertices in a graph $(m/n$ or e/v). The *vertex degree of connectivity* of a graph is the minimal number of vertices which need to be removed in order that the graph may become disconnected. A similar definition holds for the *edge degree of connectivity* (Carré, 1979, pp.142-174).

In summary, then, any network can be represented as a graph with n vertices and m edges and the corresponding graph is defined by an incidence matrix with dimensions $n \times m$. In order to illustrate the presented method, the example shown in Fig.4.4 will be used.

4.3.2 *Elements of computational hydraulics*

Flows in the open-channel networks treated in this text are supposed to be described by the de Saint Venant equations for one-dimensional free-surface flow:

Continuity: $$\frac{\partial Q}{\partial x} + b_s \frac{\partial h}{\partial t} = 0 \qquad (4.1)$$

Momentum: $$\frac{\partial Q}{\partial t} + \frac{\partial}{\partial x}\left(\beta \frac{Q^2}{A}\right) + gA\left(\frac{\partial h}{\partial x} - I_b\right) + \frac{g\,Q|Q|}{C^2 AR} = \qquad (4.2)$$

where:

Q	- discharge	I_b	- bottom slope
h	- water depth	R	- hydraulic radius
b_s	- storage width	β	- Boussinesq coefficient
A	- cross-section area	C	- Chezy resistance coefficient

Solutions of the flow conditions are obtained in the case considered here by the finite-difference method. Although all the concepts introduced are invariant with respect to the choice of the finite-difference scheme between the two most commonly used schemes, namely those of Preissmann and Abbott-Ionescu, in this chapter the applications to the Abbott-Ionescu scheme are presented (Abbott and Basco, 1989, pp.259-264). For more details see Appendix 4.1. The same methods apply to simplified formulations, but the algorithms elaborated later, although suitable for the parabolic wave approximation, will need to be modified for the kinematic wave formulation. As the compatibility conditions that have to be applied at vertices of the channel-network, equal water levels are considered due to the slightly easier treatment. The approach presented here can be easily extended to the more accurate energy-level equivalence at vertices, as well as to other implicit finite-difference schemes (Cunge *et al*, 1980, pp.111-112).

At each vertex of the graph there is an unknown denoted by $Y(i)$, $i = 1$ to n, representing the water levels at the vertices. Along each edge there is a set of unknowns denoted by $X(k,j)$, $k = 1$ to m, $j = 1$ to the number of grid points along the kth edge (more usually denoted by jj), representing alternatively discharges and water levels along the edge.

The equations to be solved can be divided into two types: first, the continuity equations at all the vertices, which have the form:

$$\sum_{ee=1}^{d_i} X\left(e(ee),\alpha \right) = 0 \qquad (4.3)$$

where:

i	- number of the vertex;
d_i	- degree of the ith vertex;

ee	- indices of all the edges incident to this vertex, $ee=1$ to d_i;
e(ee)	- numbers of edges incident to the ith vertex;
α	- 0 or $jj+1$ depending upon whether the edge is incoming or outgoing;
$X(e(ee),\alpha)$	- discharge at the beginning (or end) of the edge number *e(ee)*[1]

and, second, the system of equations along the edges which have the form:

$$\begin{bmatrix} * & * & * & & & \\ & * & * & * & & \\ & & * & * & * & \\ & & & * & * & * \\ & & & & \cdot & \cdot & \cdot \end{bmatrix} \times \begin{bmatrix} Y(i) \\ X(e,1) \\ X(e,2) \\ X(e,3) \\ \cdot \end{bmatrix} = \begin{bmatrix} * \\ * \\ * \\ * \\ \cdot \end{bmatrix} \tag{4.4}$$

with:

i	- the number of the incident vertex
e	- the number of the edge

It is seen that each of the equations from the above-presented system of equations has the general form:

$$A(e,j)\cdot X(e,j-1) + B(e,j)\cdot X(e,j) + C(e,j)\cdot X(e,j+1) \tag{4.5}$$

Beside the network topology, given by the graph in Fig.4.4, the characteristics of each channel are to be given, together, of course, with the boundary conditions. The boundary conditions introduced in this model are those of water level as a function of time and discharge as a function of time. All the boundary conditions are situated at vertices. The discharge boundary condition can be given at any vertex and if it exists it is included in the continuity equation (4.3) for that vertex. The water level boundary conditions are allowed only in the end-vertices (dendritic ends of the graph) and they are used to initialise the elimination sweep along the incident edge. This kind of boundary condition in other vertices is avoided as it would influence the definition of the control mechanism. However, if there is a need for such a boundary condition it can always be introduced by a decomposition of the given vertex into the set of end vertices (in a similar manner as explained on pages 4:16-4:17) with the same

[1] Due to the use of staggered grid in the Abbott-Ionescu scheme, the first discharge along the edge is $X(e,1)$ and it is placed at distance Δx from the vertex. In order to get the value of the discharge placed at the very beginning of the edge ($X(e,0)$), an additional continuity equation has been introduced. The same procedure has been used for the discharge at the very end of each channel.

boundary condition defined in each of them. The boundary conditions are discussed in more detail in Appendix.4.1.

4.3.3 *Systems of equations to be solved*

The system of equations concerned is composed from various sources. Each channel of the network contributes a system of linear equations (4.4) characterised by a banded matrix (tri- or penta- diagonal, dependent on the finite difference scheme used). Beside these, at each and every node of the network a continuity equation (4.3) and a compatibility condition are given. Together with the system of boundary conditions, these form one system of linear equations. This full system of equations can then be written in the form:

$$Ax = b$$

In an efficient computer code, the matrix of this system of equations (A) is never assembled. In order to economise on memory, only the non-zero entries are kept, and these are maintained in separate blocks. However, if such a matrix were to be assembled, it could always be reduced to the following form:

$$A = \begin{bmatrix} A_{11} & & & & A_{1N} \\ & A_{21} & & & A_{2N} \\ & & A_{31} & & A_{3N} \\ & & & \cdot & \cdot \\ A_{N1} & A_{N2} & A_{N3} & & A_{NN} \end{bmatrix} \tag{4.6}$$

with A_{ij} being the block matrices. The submatrices

$$\begin{bmatrix} \cdot & \cdot & A_{ii} & \cdot & \cdot & A_{iN} \end{bmatrix}$$

originate from the systems of equations connected with each of the channels (4.4) involved while the submatrix

$$\begin{bmatrix} A_{N1} & A_{N2} & A_{N3} & \cdot & \cdot & A_{NN} \end{bmatrix}$$

comes from the system of nodal continuity equations (4.3).

This particular form of matrix is usually called the *doubly bounded block diagonal form* (*DBBDF*) in the field of *sparse matrices*. It is one of the forms that are desirable for the efficient reduction of matrices (Tewarson, 1971) and it can, for example, be obtained by an algorithm called *one-way dissection* (George, 1980) or by some other partitioning algorithm (Duff *et al*, 1986, pp.239-261).

4.4 Solution methods

4.4.1 *Block Gauss elimination*

Once its matrix is reduced to this specific form, the system of equations can be solved by *block Gauss elimination* (Duff *et al*, 1986, pp.161-163) which can be represented for this form by the following algorithm:

$$D_i = A_{ii}, \qquad i = 1, 2, \ldots N-1$$

$$D_N = A_{NN} - \sum_{j=1}^{N-1} A_{Nj} D_j^{-1} A_{jN}$$

$$c_i = b_i, \qquad i = 1, 2, \ldots N-1$$

$$c_N = b_N - \sum_{j=1}^{N-1} A_{Nj} D_j^{-1} c_j$$

$$x_N = D_N^{-1} c_N$$

$$x_j = D_i^{-1} (c_i - A_{iN} c_N)$$

This method, although taking account of the advantages provided by the specific form of the matrix, does not achieve the most efficient results in cases where the size of the block-matrices is considerable. In the case of modelling channel networks this means that it becomes less efficient when the number of discretisation points used along each channel and/or the number of nodes in the network are considerable.

4.4.2 *Network algorithms*

The branched networks (with no loops involved) are presented by dendritic graphs that are commonly called trees. (See Fig.4.1a). The matrix of the system of equations belonging to any branched network can be solved without any fill-in (Parter, 1961) by different methods including the extension of the double-sweep algorithm (Cunge *et al*, 1978, pp.109-113) usually applied to branched channel networks. This method, usually referred to as a *branched network algorithm*, uses network graph topology to establish the order in which channels have to be tackled so that no additional fill-in is required.

The looped networks are characterised by graphs with loops, as represented, for example by Fig.4.1b. The solution algorithm called the *looped network algorithm* is a classical way of solving flow conditions in such networks. The method consists of two steps. The first step is the so called *local elimination,* usually performed on the matrix of each sub-system of equations corresponding to each channel represented by equation (4.4) and/or by each of the blocks A_{ii} in equation (4.6). Through the local elimination, the matrix of the systems of equations for each channel passes through the following transformation:

$$\begin{bmatrix} * & * & * & & & & \\ & * & * & * & & & \\ & & * & * & * & & \\ & & & \cdot & \cdot & \cdot & \\ & & & & * & * & * \\ & & & & & * & * & * \end{bmatrix} \quad \text{goes over into} \quad \begin{bmatrix} * & 1 & & & & & * \\ * & & 1 & & & & * \\ * & & & 1 & & & * \\ \cdot & & & & \cdot & & \cdot \\ * & & & & & 1 & * \\ * & & & & & & 1 & * \end{bmatrix}$$

Upon local elimination, all the first and last equations from these systems of equations are used in order to substitute values for the discharges in the nodal equations. The solution of these systems of nodal equations constitutes a second part of the looped network algorithm. This system of nodal equations has an order equal to the number of nodes and is expressed in terms of unknowns related to the nodes. Its matrix, although banded, has a significant band-width but it is at the same time very sparse. For the solution of this system of equations, different direct and iterative methods can be employed and these are indeed often used in currently available commercial software for the solution of flows in open channel networks.

4.4.3 *Other direct methods for sparse matrices*

The iterative methods, although preferred by some authors, are not considered in any detail here due to their lack of transparency and often strong dependence on the numerical values. The field of *sparse matrix technology* is developing rapidly, however, with the development of linear programming and numerical methods which produce huge systems of linear equations characterised by very sparse matrices of the systems (Duff *et al*, 1986, p.14). Some of these methods seem quite interesting for possible application in the field of channel networks.

Most of these algorithms are closely related to certain applications of graph theory. They may be introduced through the definition of the *elimination graph* associated with the sparse matrix: for any $n \times n$ (i.e. square) matrix the associated elimination graph is a graph with n nodes with the property that if a_{ij} is a non-zero entry in a matrix then the nodes i and j of the graph are connected. Symmetric matrices are usually represented by undirected graphs while unsymmetric ones utilise di-graphs.

In our case, the structure of the matrix is sometimes symmetric (dependant on the finite difference scheme used) but the values of the entries are surely not. However, due to its structure, this matrix can still be considered as a 'nearly-symmetric' one.

For example, the *minimal degree algorithm* (Tinney, 1967) owes even its name to a property of the corresponding elimination graphs. This algorithm is suitable for the solution of symmetric matrices while its counterpart for unsymmetric matrices is provided by the so called Markovitcz criteria (Duff *et al*, 1986, p.128). The minimal degree algorithm, as well as the Markowicz criteria, are based on the criterion of minimal *fill-in* (Duff *et al*, 1986, p.129). That is achieved, at least locally, if the pivot row has the smallest number of non-zero entries. Translated into the world of graph representation, this means that the node with the minimal degree is being eliminated first (as indeed the name of the method again indicates).

This method, when applied to the channel network, results in a solution algorithm which, at least in the first part, coincides with the classical looped network algorithm (i.e. local elimination); when applied for solution of nodal equations it seems to be very powerful until the minimal degree in the elimination graph reaches a value of about four. If applied further, so much additional 'fill -in' is introduced that the purpose of using a sparse matrix algorithm at all is undermined (Stelling,G.S., private communication).

The application of methods such as a *one way dissection* (George, 1980) or a *nested dissection* (George, 1973) or *node tearing* (Duff *et al*, 1986, pp.254-261) results in a partitioning of the matrix in *DBBD* form, and making possible the solution of the system by some specific economical method (i.e. Block Gauss Elimination). The choice of the separator set does influence the size of the block A_{NN} and consequently the computational effort involved in the solution. The smaller separator set reduces drastically the computational effort (Duff *et al*, 1986, p.258). Hence, the successful dissection or node tearing is the one which minimises the separator set and still maintains the separation of the other vertices. When applied to the hydrodynamic network problems, all of these methods result in a separator set being a set of all nodes of the network graph and consequently it leads to the matrix being partitioned into its *DBBD* form as given in equation (4.6).

The method of Wang (1981) for the solution of banded matrices on parallel processors has some very interesting features in this connection. Wang partitions the matrix into *p* blocks of size *k* and performs elimination and substitution separately for each block (Duff *et al*, 1986, pp.208-210). However, this introduces some fill-in at boundary columns of the blocks in order to facilitate parallel computation, while the double-sweep algorithm, which is highly recursive, is not suitable for parallel computation. Wang's partitioning is driven by the size of the problem and for the purpose of solving it using parallel processing. If this partitioning is done in such a way that each block is reduced to one channel, then we are back at the looped network algorithm. However, if partitioning is done in some other ways, then it becomes interesting for applications on network problems.

4.5 The generalised solution algorithm

The generalised network solution algorithm is an adaptive algorithm. It consists of two main parts: the definition of the control mechanism (being the adaptation transducer) and the actual calculation of the flow conditions. The control mechanism required to steer the calculation procedure is obtained at the adaptation phase of the general algorithm which is performed before the actual time-dependant calculation starts. In effect, the solution algorithm is being tuned to best suit it to the problem that it is going to solve. The logical operator which defines the control mechanism is based on the network topology and it utilises the graph-theoretical approach introduced earlier in this Chapter. The sequence of edges to be followed in the elimination sweep and some other values used for control of the solution procedure are defined through a set of algorithms performed on the incidence matrix of the graph.

For a certain fixed network topology, this procedure does not need to be repeated during the calculation, so that it can be performed as a 'pre-conditioning' process, or 'pre-process'. In the field of sparse matrix technology, a similar feature is usually associated with the so called *static data structure* (Pisanetsky, 1984, pp.28-29) which consists of two parts: the symbolic one performed only once, and the numeric one which is usually repeated. The full advantage of introducing such systems is seen in solutions of initial value problems where a lot of systems of linear equations with the same structure, but different values, have to be solved.

The calculation of the flow conditions itself consists of two parts, the one for flow conditions in the dendritic part of the graph and the other for flow conditions in the cyclic part of the graph. A considerable gain in performance is obtained by splitting such a network into components which are either entirely branched or entirely looped. This makes possible the application of the particularly simple branched algorithm to the branched components, with a separate algorithm applied to the rest of the network. Therefore, in the definition of the control mechanism (adaptation phase), the *splitting algorithm* splits the graph into its dendritic and cyclic parts. On the dendritic part of the graph, the standard branched network algorithm is applied, but with the only difference that the whole algorithm for the solution of the cyclic part is placed between the elimination sweep and the substitution sweep.

The ordering of the generalised channel network solution algorithm can be schematised for structured programming purposes as shown in the following table:

ADAPTATION PHASE: - splitting up dendritic and cyclic parts of network - definition of sequence for global elimination method - orientation of the graph	
loop on all the time steeps	
	CALCULATION OF A, B, C, AND D, COEFFICIENTS FOR ALL THE EDGES OF THE NETWORK
	FIRST PART OF THE BRANCHED ALGORITHM: elimination sweep on all the dendritic parts of the network
	GLOBAL ELIMINATION ALGORITHM ON THE CYCLIC PART OF THE NETWORK: - elimination sweep - intermediate sweep - substitution sweep
	SECOND PART OF THE BRANCHED ALGORITHM: substitution sweep on all the dendritic parts of the network

The flow conditions in the cyclic part of the graph are solved by a *global elimination algorithm*, which is meant as a replacement for the commonly used looped-channel network algorithm. The governing idea is to circumvent altogether the phase in the standard looped algorithm where a system of n equations with n unknowns is formed (with n being the number of vertices in the network).

The global elimination method is based on a combination of automatised symbolic calculation and the use of the branched network algorithm. In the other words, it is based on the Gauss-elimination that is performed in any case on each of the channels within the local elimination (Volkov, 1986, pp.155-160). Hence, in some respect it follows the algorithmic structure required for the direct solution of one-dimensional flow, a feature which seems to be very difficult to implement when using iterative methods. The whole global elimination algorithm is again in the form of a symbolic computation, in which mathematical operations are performed on symbols as if they were numbers (e.g. Wolfram, 1988).

In order to apply the global elimination method, a set of *symbols* as well as the corresponding sequences of edges have to be defined. This is done within the adaptation phase of the generalised solution algorithm. This algorithm is called a *decomposition algorithm* and it is performed on the incidence matrix of the cyclic part of the graph.

4.5.1 *Adaptation phase - the definition of the control mechanism*

4.5.1.1 The splitting algorithm

Although some standard elements of graph theory can be employed in order to develop an algorithm for splitting the channel network into dendritic and cyclic components, the precise problem which arises here is not treated in standard algorithmic graph theory so that it became necessary to develop an algorithm especially for this purpose.

The principal innovation is to use the notion of *vertex degree*, defined as the number of edges incident to a vertex. All the vertices at the ends (or beginnings, depending on the definition) of the dendritic parts of the graph have a degree of unity. Using this criteria, they can easily be detected from the incidence matrix of the graph, where the sum of all the elements in each row represents the degree of the corresponding vertex.

The whole algorithm is performed on a sequence of matrices (each element of which is denoted as a C-matrix) which is initiated by the incidence matrix and the final form of which contains only the cyclic part of the graph. Four additional arrays are used for constructing the sequence; the first are called a 'DE-list' and a 'DV-list' and contain respectively the numbers of the edges and the numbers of vertices involved in the dendritic part of the graph, and the second pair are called a 'CE-list' and a 'CV-list' and contain the numbers of the edges and vertices respectively involved in the cyclic part of the graph.

The algorithm is initiated by a search procedure conducted through all the rows of the initial C-matrix (the incidence matrix) for vertices with a degree of one. When such a vertex is detected, its number is placed in the DV-list while the only edge incident to it is removed from the original C-matrix and its number is placed in the DE-list. The procedure is repeated, in the whole loop and through all the rows of the matrix, until no further vertex of degree unity is detected. Finally, from the non-zero entries of the C-matrix, the CV-list and CE-list with vertices and edges involved in the cyclic part are constructed while the final C-matrix is reduced to its non-zero rows and columns only. At the same time, the DE-list then contains the numbers of all the edges and the DV-list the numbers of all the vertices in the dendritic parts of the graph.

As an illustration of the operation of this algorithm, the results of its application to the simple example of the graph shown in Fig.4.3 are presented in Fig.4.4. As can be seen from Fig.4.4, this algorithm has partitioned the graph used as an example in Fig.4.3 in a manner suited to the separation of the network-solution algorithm. The complexity of the separation algorithm has not been examined in detail. However, it seems that even in the worst case it does not necessitate more than n^2 checking operations, with n being the number of vertices in a graph. As it is applied only at the time of instantiation, this is considered entirely acceptable. The details of this algorithm and its pseudo-code are given in Appendix 4.2.

4.5.1.2 The decomposition algorithm

The decomposition algorithm defines the list of symbols and the corresponding sequence of edges and vertices needed to operationalise the global elimination algorithm. The basic idea is to chose a vertex such that the value of the water level at this vertex can be carried as a symbol throughout the elimination algorithm. The elimination sweep can start from such a vertex, in any direction, in a similar manner as is done in the branched algorithm from an end vertex. Hence, a symbol-vertex can be viewed as being decomposed into as many vertices as its degree. Therefore, the role of the decomposition algorithm is to chose the symbol-vertices and consequently define all the 'dendritic' paths[2] originating from the decomposed symbol-vertex. The number of symbol-vertices to be decomposed depends on the complexity of the graph interconnections.

C matrix

$$\begin{bmatrix} 1&0&1&0&0&0&0&0&0&0&0&1\\ 1&1&0&0&0&0&0&0&0&0&0&0\\ 0&1&1&1&0&0&0&1&0&0&0&0\\ 0&0&0&1&1&0&0&0&0&0&0&0\\ 0&0&0&0&1&1&0&0&0&0&0&0\\ 0&0&0&0&0&1&1&0&1&0&0&0\\ 0&0&0&0&0&0&1&1&0&0&1&0\\ 0&0&0&0&0&0&0&0&1&1&0&0\\ 0&0&0&0&0&0&0&0&0&1&1&1 \end{bmatrix}$$

DE list	DV list	CE list	CV list
5	5	1	1
6	6	2	2
10	10	3	3
11	11	7	7
17	15	8	8
4	4	9	9
		12	12
		13	13
		14	14
		15	
		16	
		18	

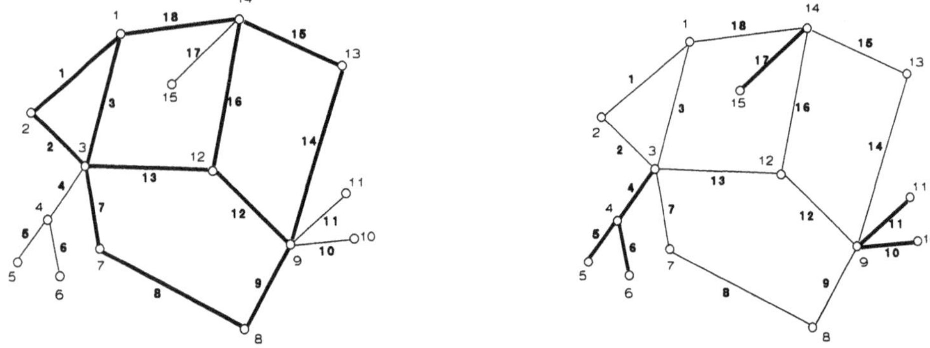

a)Cyclic part of the graph b) Dendritic part of the graph

Fig.4.5

[2] A *path* is a sequence of alternating edges and vertices.

The presented decomposition algorithm is not unique. The criterion for its choice was to have the smallest number of symbols which must be carried through the elimination procedure. Accordingly the vertices with the highest degree are chosen to be decomposed first. Another possibility would be to organise the elimination sweep according to the loops of the graph, but for that purpose an algorithm for loop generation would be required (Osiadacz, 1987, pp.50-68). In general, the best way to chose the decomposition criteria would be to pose an optimisation problem with respect to the overall number of operations in the elimination algorithm. In that respect it is important to notice that the number of sweeps to which each edge is exposed increases with the number of symbols. Edges corresponding to the first symbol have only three sweeps, while for each successive symbol the number of sweeps increases by unity.

The decomposition algorithm is performed on a working matrix (WM) initiated as a copy of the incidence matrix of the cyclic graph, which is the final C-matrix obtained from the splitting algorithm. The results of the decomposition algorithm are: the number of symbols, NS, the array S(NS) with the symbol-vertices' numbers, the array NP(NS) with numbers of 'dendritic' paths originated by the decomposition of the corresponding symbol-vertices, the array LP(NS,NP(NS)) with the lengths of each of the 'dendritic' paths[3], the array E(NS,NP(NS),LP(NS,NP(NS))) with the edges' numbers of all the edges involved in each 'dendritic' path, and the array V(NS,NP(NS), LP(NS,NP(NS))-1) with the vertices' numbers of all the vertices involved in each 'dendritic' path.

The decomposition algorithm first calculates the degrees of the vertices of the WM. The vertex with the highest degree, or any one of them if there are more vertices with the same degree, becomes the symbol-vertex, whereupon it is removed from the working matrix (WM) and placed in the list of symbols (S), and the number of symbols (NS) is increased by one. Further, the algorithm searches for a column of the working matrix with only one entry, it increases the number of paths (NP) originating from the removed vertex by one, it places the edge number corresponding to that column in the list of edges (E) and it sets the length of that path (LP) to unity. After this, the algorithm searches for a dendritic path following that edge. For each successive dendritic edge, the vertex number is placed in the list of vertices (V), the length of the path (LP) is increased by one and the edge number is placed in the list of edges (E). The algorithm then searches for another column with a single entry and repeats the whole procedure. If no column contains a single entry and the WM-matrix is not completely zero, then the procedure starts again from the point where the calculation of the degrees of all the vertices is performed. The procedure is repeated until all the entries of the WM are equal to zero.

The array E(k,m,n) partitions the set of all the edges of a graph with respect to the index k. A subset of the set of all the edges whose numbers are placed in the array E with the first index equal to k will be called a k-subset. In the global elimination algorithm, for each of these subsets a different number of symbols will be involved. The number of symbols for each subset is equal to its index k. For the first subset, the symbol will be the water level (Y) at the vertex

---

[3] The *length of a path* is measured by the number of edges traversed.

whose number is the first element from the array S. For the second subset, the symbols will be the water levels in vertices whose numbers are the first and the second element in the array S. Thus the symbols corresponding to the k-subset are $Y(S(i))$, $i = 1$ to k. In the same manner, the subsets corresponding to a symbol $Y(S(i))$ are the k-subsets, $k = i$ to NS.

By way of an example, when applied to the graph exemplified in Fig.4.5a, the above algorithms provide the results shown in Fig.4.6. For details see Appendix 4.2.

-number of symbols **NS** = 2,

-numbers of symbol-vertices
S = {3,9},

-number of paths corresponding to each symbol

NP = { 4 , 1 },

-length of each path **LP**={{2,2,3,1},{4}},

-list of vertices per each symbol and path

V={{{2},{1},{7,8},∅} ,{{12,14,13}},

-list of edges per each symbol and path
E={{{2,1},{3,18},{7,8,9},{13}},{{12,16,15,14}}

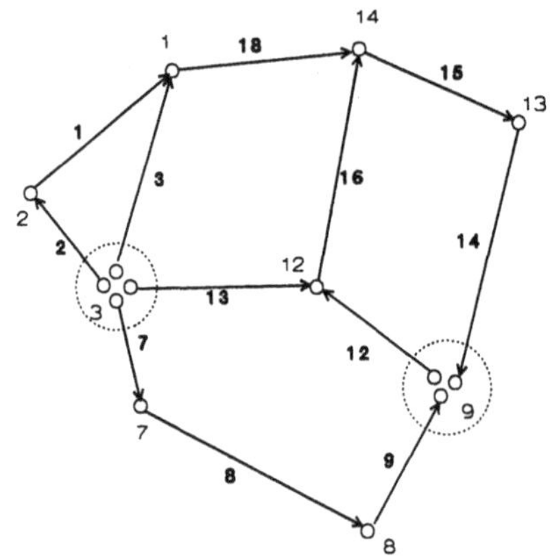

Fig.4.6

4.5.1.3 The orientation algorithm

The last of the algorithms used in the definition of the control mechanism is the orientation algorithm. In this algorithm the graph is transformed into an *orientated graph (or digraph*; see Chen, 1990, pp.75-81). This transformation again starts out from the incidence matrix of the graph. The physical basis of this procedure is that the ordering obtained at every stage in every algorithm should correspond to the flow of information through the physical system as defined by its characteristic structure (Abbott and Basco, 1989, p.210).

Within the solution procedure for a single channel, it is assumed that the positive direction of information flow has already been related to the increase in the grid-point numeration: see Fig.6. Each channel in the network is treated in this way. Accordingly, an orientation has to be assigned to each edge of the graph representing the positive direction of information flow.

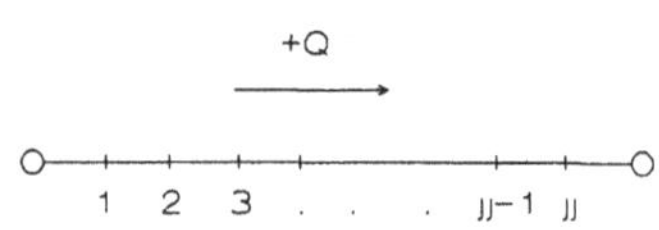

Fig.4.7

As the assumption of a positive direction of information flow in an edge corresponds to an increase in the grid index, the notion arises of orientating all the edges in the direction in which they will be transversed by the elimination sweep i.e. following the defined sequence of edges and vertices. The discretisation of edges in this way simplifies the elimination algorithm, as each edge is 'attacked', so to say, from its 'beginning'. The whole calculation procedure will henceforth be performed according to this diagraph. It is important to notice that each row of the new incidence matrix provides the mass or energy continuity equation concerned directly at the corresponding vertex.

Upon application of this algorithm to the graph exemplified in Fig.4.4, the resulting diagraph can be presented in the incidence matrix and the diagraph presented in Fig.4.8. The difference between the graphs presented in Fig.4.4 and Fig.4.8 is that the orientation of the edges presented in the diagraph of Fig.4.8 accords with the defined control mechanism. For more details see, Appendix 4.2.

The incidence matrix

$$
\begin{bmatrix}
1 & 0 & 1 & 0 & 0 & 0 & 0 & 0 & 0 & 0 & 0 & 0 & 0 & 0 & 0 & 0 & 0 & -1 \\
-1 & 1 & 0 & 0 & 0 & 0 & 0 & 0 & 0 & 0 & 0 & 0 & 0 & 0 & 0 & 0 & 0 & 0 \\
0 & -1 & -1 & 1 & 0 & 0 & -1 & 0 & 0 & 0 & 0 & 0 & -1 & 0 & 0 & 0 & 0 & 0 \\
0 & 0 & 0 & -1 & 1 & 1 & 0 & 0 & 0 & 0 & 0 & 0 & 0 & 0 & 0 & 0 & 0 & 0 \\
0 & 0 & 0 & 0 & -1 & 0 & 0 & 0 & 0 & 0 & 0 & 0 & 0 & 0 & 0 & 0 & 0 & 0 \\
0 & 0 & 0 & 0 & 0 & -1 & 0 & 0 & 0 & 0 & 0 & 0 & 0 & 0 & 0 & 0 & 0 & 0 \\
0 & 0 & 0 & 0 & 0 & 0 & 1 & -1 & 0 & 0 & 0 & 0 & 0 & 0 & 0 & 0 & 0 & 0 \\
0 & 0 & 0 & 0 & 0 & 0 & 0 & 1 & -1 & 0 & 0 & 0 & 0 & 0 & 0 & 0 & 0 & 0 \\
0 & 0 & 0 & 0 & 0 & 0 & 0 & 0 & 1 & 1 & 1 & -1 & 0 & 1 & 0 & 0 & 0 & 0 \\
0 & 0 & 0 & 0 & 0 & 0 & 0 & 0 & 0 & -1 & 0 & 0 & 0 & 0 & 0 & 0 & 0 & 0 \\
0 & 0 & 0 & 0 & 0 & 0 & 0 & 0 & 0 & 0 & -1 & 0 & 0 & 0 & 0 & 0 & 0 & 0 \\
0 & 0 & 0 & 0 & 0 & 0 & 0 & 0 & 0 & 0 & 0 & 1 & 1 & 0 & 0 & -1 & 0 & 0 \\
0 & 0 & 0 & 0 & 0 & 0 & 0 & 0 & 0 & 0 & 0 & 0 & 0 & -1 & 1 & 0 & 0 & 0 \\
0 & 0 & 0 & 0 & 0 & 0 & 0 & 0 & 0 & 0 & 0 & 0 & 0 & 0 & -1 & 1 & 1 & 1 \\
0 & 0 & 0 & 0 & 0 & 0 & 0 & 0 & 0 & 0 & 0 & 0 & 0 & 0 & 0 & 0 & -1 & 0
\end{bmatrix}
$$

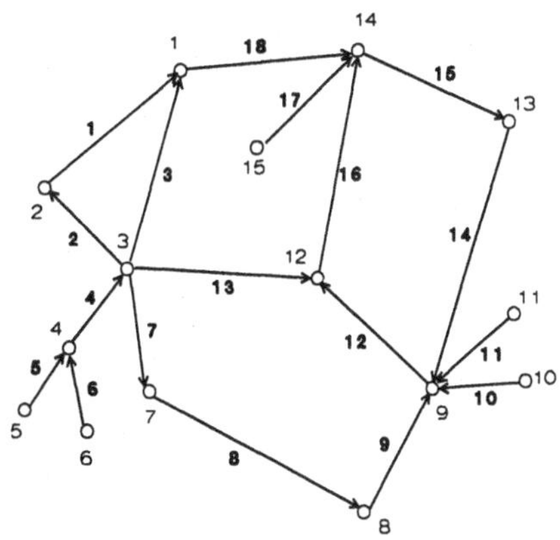

Fig.4.8

4.5.2 *The solution algorithm*

4.5.2.1 The dendritic part: the branched network algorithm

As the dendritic part of the network is solved by the classical branched network algorithm (Cunge *et al*, 1980, pp.109-113), for the elimination sweep, within each edge, the usual expressions for the recurrence relations are used. The sequence of edges in the elimination sweep is controlled by the DE-list and the DV-list of dendritic edges and vertices respectively, as defined by the splitting algorithm. Details of the solution algorithm are exposed in Appendix 4.3.

If the network graph is a tree, a graph with no loops, then upon completion of the elimination sweep the continuity equation at the last vertex is performed followed by the substitution sweep on all the dendritic edges. If there is a cyclic part in the graph, then upon completion of the elimination sweep on the dendritic part the whole global elimination algorithm for the solution of the cyclic part of the network is performed. Only then does the substitution sweep on all the dendritic edges continue.

4.5.2.2 The cyclic part: the global elimination algorithm

The global elimination algorithm was developed as a method for the solution of the system of equations arising from the concrete physical problem (flow in looped channel networks); hence, help from the network topology was available and use was made of this.

The idea behind this method is to transform a looped network into a 'quasi dendritic' one by decomposition of some nodes. Consequently a form of branched network algorithm (Cunge *et al*, 1980, pp.109-113) can be applied. However, since any such network is only 'quasi dendritic' its solution will produce some 'fill-in', unlike the solution of a real dendritic network. Therefore the decomposition nodes (symbols) are to be chosen carefully in order to minimise, at least approximately, the amount of 'fill-in'. (It is not strictly minimised because no optimisation problem is in fact solved.)

In some sense this can be seen as a combination of a node tearing technique and the method proposed by Wang. Similarly to the process of node tearing, it chooses a minimal separator set for partitioning of the matrix, but based on another criterion. This approach is elaborated in Appendix 4.4. The similarity with the method of Wang is in the elimination procedure, but the main difference is that the partitioned blocks correspond to parts of graph that become 'quasi dendritic' upon decomposition of one node. In such a way, the matrix is partitioned in as many blocks as the number of nodes that are decomposed.

The global elimination method can as well be viewed as a form of symbolic computation. The replacement of the value of the water level at a vertex with a symbol makes it possible to regard that vertex as if it were an end-vertex for all the incident edges. This, of course, greatly simplifies the graph interconnections. In the decomposition algorithm, as a preparation for the global elimination algorithm, this is done with the least number of vertices, so that the graph becomes quasi-dendritic. The flow conditions in the network are then solved with an algorithm based on the branched-network algorithm. The main differences are confined to changed expressions for the recurrence relations and the introduction of closure equations at the symbol-vertices to make-up for their decomposition. For a more mathematically strict description, see Appendix.4.4.

Through the development of general mathematical computational tools over the last few years (such as *Mathematica* or *Maple*), it has become possible to perform symbolic computations directly on a computer. However, the possibilities for performing the complete global elimination algorithm using mathematical tools, although very promising, still remain largely to be explored.

At the symbol-vertices which are defined in the decomposition algorithm (array S), values of the unknowns, $Y(\mathrm{S}(i))$, are assumed to be given. We start with the first of these values $Y(\mathrm{S}(1))$ and by means of symbolic calculation we perform the elimination sweep on all dendritic walks, defined as a consequence of the decomposition of that vertex (E(*1,m,n*)). This applies, as well, to the first-subset. The same procedure is repeated with all the symbol- vertices (i.e. the array S).

When this process is completed, an intermediate sweep is performed in the reverse order, first on the *k*-subset, *k* = NS, and then on the other *k*-subsets, *k* = NS-1 to 1 in steps of -1. Upon completion of an intermediate sweep on one *k*-subset, the closure equation for the symbol *Y*(S(*k*)), being the continuity equation at the decomposed vertex, is used to obtain the value of the symbol. After completing this process, a substitution sweep is performed on all the subsets corresponding to that symbol. The intermediate sweep is repeated for each of the *k*-subsets in the inverse order to the one used in the elimination sweep. The details of these process are set out in the Appendix 4.3.

4.6 Results

A program was developed (named GEM) to test the feasibility of the method introduced here. The usual static, steady and unsteady tests were successfully applied on multiple network configurations. Some layouts of the networks that were tested are presented in Fig.4.9.

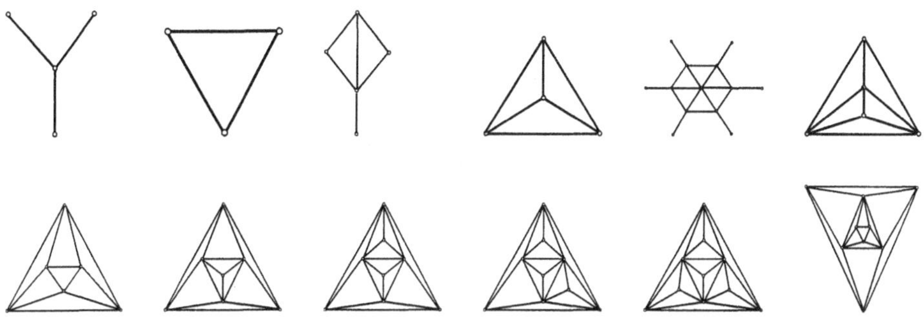

Fig.4.9

Most of the graphs from Fig.4.9 have been chosen for testing the performance of the global elimination method for the solution of the cyclic parts of networks because they are the most complex planar graphs for a given number of vertices. These were tested for a variety of inflow and outflow points to the initial network.

In order to examine the efficiency of the generalised network solution method, a few tests were made concerning the computation time. First, the instantiation time of the pre-processor of the GEM program, being largely determined by the definition of the control mechanism, was tested. For this purpose a simplified program was developed for the benchmark solution of just a single channel. It uses all the procedures needed by the GEM program but no definition of the control mechanism.

A single channel with 120 grid points was used as an example. Flow in this channel was calculated by both programs for the same number of time steps. The duration of the run of the simplified program defines the time unit while the duration of the run of the GEM program is expressed in units of their run time normalised by this time. Representing the relative time spend by the GEM program as a function of the number of time steps gives the result presented in Fig.4.10 by the single channel line. From this graph it can be seen that the relative time decreases with the increase in the number of time steps. This result was expected because the definition of the control mechanism is constructed only once.

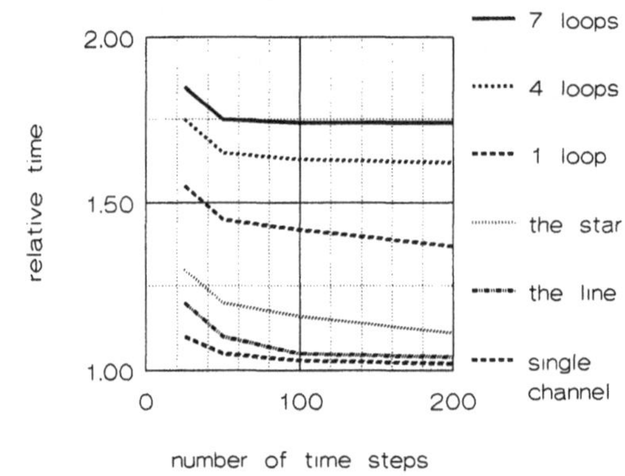

Fig.4.10

The following tests were made in order to estimate the performance of the generalised network solution method, and especially the global elimination method, on graphs with increasing complexity.

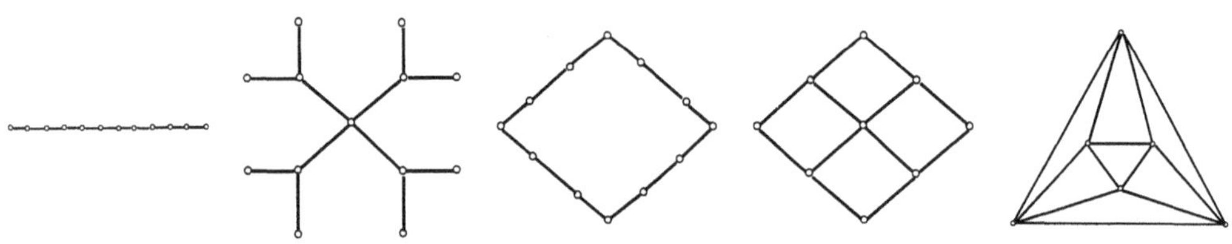

'the line'	'the star'			
loop = 0	loop = 0	loop = 1	loop = 4	loop = 7
e/v = 0.9	*e/v* = 0.9	*e/v* = 1	*e/v* = 1.3	*e/v* = 2
conec. = 1	conec. = 1	conec. = 2	conec. = 3	conec. = 4
symb. = 0	symb. = 0	symb. = 1	symb. = 2	symb. = 3

Fig.4.11

A set of graphs was chosen with an equal number of edges (12) and an equal number of grid points along each edge (10). All these graphs then have the same number (120) of grid points as the single channel used in the previous example, so that the time used by the simplified program can again be used for normalisation. The chosen set of graphs has an invariant number of edges and grid points but the complexity of the graphs increases successively. As measures of graph complexity, several numbers were used: the number of loops in a graph, the ratio between the number of edges and the number of vertices, the vertex degree of connectivity and, finally, the number of symbols required for the global elimination method. The layouts of these graphs together with their characteristic measures are presented in Fig.4.11.

The relative time used by the GEM program for the solution of each of these networks, as a function of the number of time steps, has been presented in Fig.4.10. As can be expected, the relative elapsed time increased with the complexity of the graph. However, the increase appears to be primarily concentrated in the instantiation procedure: this time does not depend upon the choice of points at which inflow and outflow are applied to the network for a given number of such points.

The results shown in Fig.4.10 can then also be presented as functions of the various measures of graph complexity. In Fig.4.12, such results are presented for runs with 200 time steps.

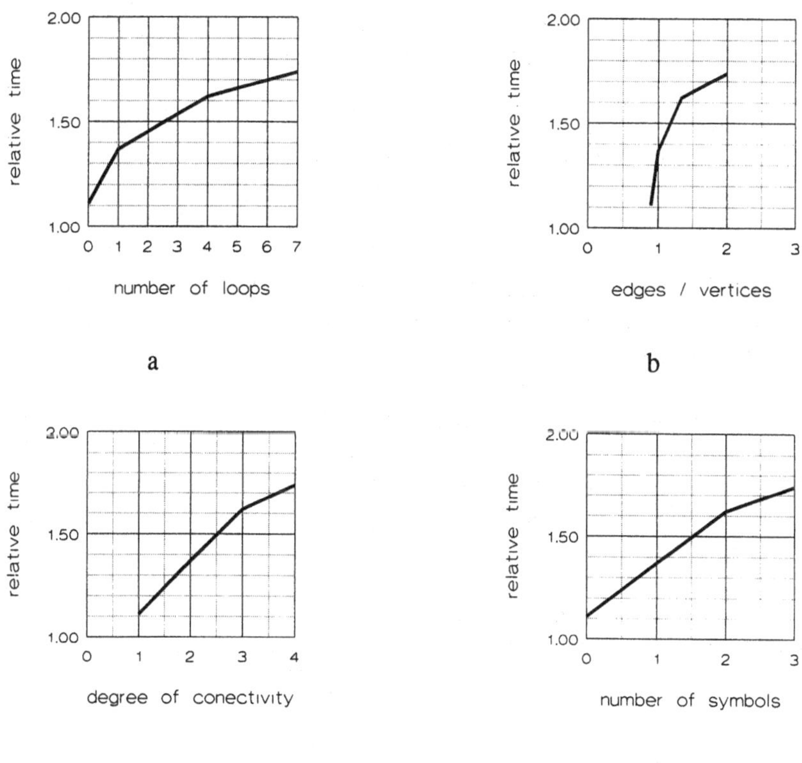

Fig.4.12

From the functions presented in Fig.4.12 it is clear that, with an increase in any of the measures of the graph complexity, the required relative time also increases. However, the trend of these functions again shows that the rate of change of the elapsed time with the increase of graph complexity actually decreases. This result is a particularly positive one for the method presented here.

At the same time, it should be observed that the time consumed by the global elimination algorithm does not depend at all on the number of vertices. The graph with seven loops has the smallest number of vertices, only six, but the highest relative time consumption. This appears to be a most unusually positive feature for an algorithm for the solution of looped networks.

4.7 Conclusions

The proposed adaptive algorithm offers an efficient way to solve unsteady flow conditions in any network. Due to the employment of an adaptive control mechanism, each of the parts of the network is solved with the most suitable method. The fact that the control mechanism is introduced entirely as a pre-conditioning process makes the increase in computational time consequent upon the multiple connectedness of the network negligible, especially for the longer simulations that are usually required. Thus, although the running time does increase with the graph complexity, the actual rate of increase is itself reduced with increasing complexity.

Although the time consumption of the algorithm, unlike the classical looped network algorithms, does not increase markedly with the increase in the number of vertices, it is of course still influenced by the overall number of grid points in the same manner as a single channel solution. Altogether it can be concluded that the global elimination algorithm based on symbolic computation has proved to be feasible as a method for the practical solution of flow conditions in looped networks. However, the global elimination method can as well be used for solution of the nearly banded matrices originating from other fields provided that they fulfil the same conditions.

It is interesting to point out that the solution of the dendritic part of the graph, being the well known branched network algorithm based on the double-sweep algorithm, is just a trivial case of the global elimination algorithm. If the number of symbols is equal to zero (i.e for dendritic networks), and consequently the set of all symbols is void, the global elimination algorithm reduces to the branched network algorithm.

Although, the method has demonstrated its applicability, there are still several points requiring further investigation. One of these is the choice of the criterion for the decomposition algorithm. The introduction of an optimisation problem for the minimisation of the number of operations in the global elimination algorithm would almost certainly provide better criteria then the one in current use. A further matter for investigation is the implementation of an edge under supercritical flow conditions that are not modified to maintain two directions of information flow, or a change in the network topology due to the drying-out of channels that are not underslotted. Due to the fact that the global elimination algorithm follows the algorithmic structure required for single channel solution, it is anticipated that these extensions will be feasible.

CHAPTER 5

ADI+ algorithm for the solution of two-dimensional nearly horizontal flow

5.1 Introduction

Two-dimensional nearly horizontal flow is an approximation of general three-dimensional flow that is commonly used in physical situations characterised by free surfaces where the assumption of hydrostatic pressure distribution over the vertical can be justified. In hydraulics, the most commonly encountered cases of such flows are flows in open channels and tidal flows. Due to the reduction of the physical space to just two dimensions, the flow is in these cases described by its depth and horizontal velocity which is averaged in the vertical direction and usually taken along two orthogonal components orientated in the directions of the main coordinate directions.

The set of continuum equations describing two-dimensional nearly horizontal flow consists of three elements (Abbott *et al*, 1973; Weiyan, 1992, p.38); in hydraulics these are most commonly the equation of conservation of mass and the equations of conservation of momentum usually taken in two orthogonal directions. In conservative form, these equations can be written as:

$$\frac{\partial f}{\partial t} + \frac{\partial g_1}{\partial x_1} + \frac{\partial g_2}{\partial x_2} = 0 \tag{5.1}$$

with:

$$f = \begin{bmatrix} h \\ u_1 h \\ u_2 h \end{bmatrix}, \quad g_1 = \begin{bmatrix} u_1 h \\ u_1^2 h + \dfrac{g h^2}{2} \\ u_1 u_2 h \end{bmatrix}, \quad g_2 = \begin{bmatrix} u_2 h \\ u_2 u_1 h \\ u_2^2 h + \dfrac{g h^2}{2} \end{bmatrix}$$

and with:

x_1,x_2	- space coordinates	u_1	- velocity in x_1 direction
t	- time coordinate	u_2	- velocity in x_2 direction
h	- water depth	g	- acceleration due to gravity

5.2 Standard ADI algorithm

A now rather standard algorithm for the solution of two-dimensional nearly horizontal flow by the method of finite differences is the so called alternating direction algorithm (ADI) (Abbott *et al*, 1973; Abbott *et al*, 1981; Stelling *et al*, 1986; DHI, 1992). Its main characteristic is the splitting of the calculation into two series of 'one-dimensional' calculations which are mutually orthogonal. For this purpose, the time step is divided into two parts and the system of governing equations is solved in steps. By this splitting of the two-dimensional problem into a series of 'one-dimensional' ones, a considerable reduction in computational effort is obtained in comparison with the methods treating such a problem in two dimensions simultaneously.

In the ADI algorithm the solution procedure is split in such a way that in one direction the conservation of mass and the conservation of the direction-corresponding momentum are solved, and, after that, in the other direction, the conservation of mass is again solved but now with the conservation of momentum introduced in that direction. A part of the typical grid for the ADI algorithm is presented in Fig.5.1.

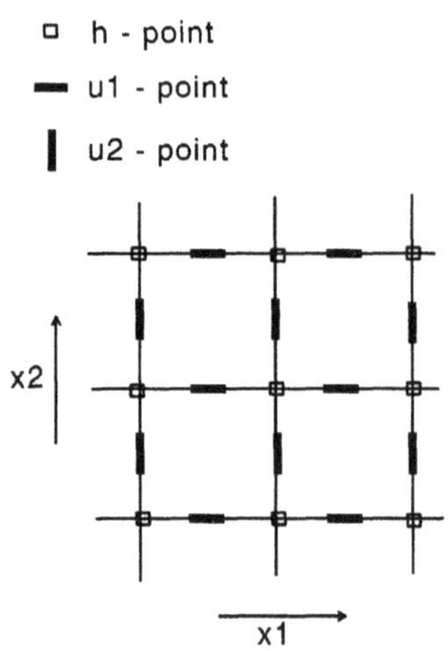

Fig.5.1

For the purpose of the development of a flow adaptive scheme, the simplest form of the shallow water equations, without convective terms, was used:

$$\frac{\partial h}{\partial t} + \bar{h}\left(\frac{\partial u_1}{\partial x_1} + \frac{\partial u_2}{\partial x_2}\right) = 0 \tag{5.2}$$

$$\frac{\partial u_1}{\partial t} + g\frac{\partial h}{\partial x_1} = 0 \tag{5.3}$$

$$\frac{\partial u_2}{\partial t} + g\frac{\partial h}{\partial x_2} = 0 \tag{5.4}$$

Details of the solutions of these equations using the ADI algorithm are given in Appendix 5.1: a model based on that algorithm was used to provide reference results for the purposes of comparison.

The introduction of the convective terms necessarily involves more points from other grid lines than just the one considered at any one time. One possibility is that these terms are approximated in an explicit manner, which brings in stability limitations on the time steps used. In order to off-set this effect, the approximation of convective terms in an 'implicit manner' is introduced, usually using the technique called 'side feeding' (Abbott, 1979, pp.211). It comprises the use of already-calculated values from the upper time level, first taken from the one side of the operator and then from the other side, which makes the terms implicit . In order to centre the effects of side feeding, the time step is then usually split into four instead of two parts, so that within each time step two different directions of advancement of the double sweeps are employed and this 'centres' the effects of 'side feeding' over the whole cycle of the NO.

5.3 Why a new algorithm?

All this, and other developments besides, make the ADI algorithms efficient and accurate, and hence very widely used. However, the ADI algorithm, although unconditionally stable (Abbott *et al*, 1981; Abbott and Basco, 1989, pp.270-272), still introduces limitations on the Courant numbers that are acceptable due to accuracy requirements (Weare, 1979). The use of exaggerated time steps can cause considerable errors in the propagation velocity through narrow channels and immediately behind islands (even if the grid is laid in the same direction as the channel) due to the effect of superimposing calculations arising from different directions: the information within one time step can advance only along one of the grid lines, or at most make only one 'turn' into the other direction during a single time step.

Even bigger problems arise when there is a 'narrow' channel laid down at an angle, and specifically at 45° to one of the main directions (x_1 or x_2: See Fig.5.2.) The 'narrowness' of the channel is measured in the number of grid points in each direction. In such a case the resulting velocities of the waves advancing through such a channel are highly reduced, a situation which does not of course occur in nature.

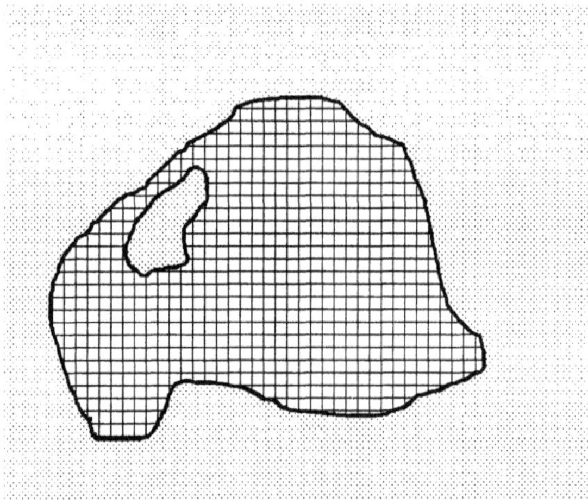

Fig.5.2

If the channel is very narrow (see, as an example, Fig.5.3), it can even happen that the information does not advance through it at all. In order to avoid these problems, usually a finer grid is used in the whole domain or the refinement of the grid is localised to the problematic area only - or a local one-dimensional model may be used. However, there are other possibilities for adaptivity which might improve the course of the computations in such cases.

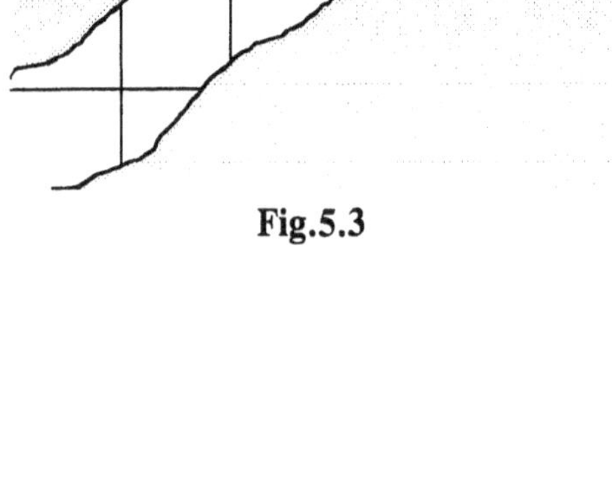

Fig.5.3

It seems that the cause for this discrepancy is the specific feature of ADI algorithms, of splitting the calculation into two perpendicular directions, which is in this case unfavourable. We can, however, locate the origin of the whole problem at a deeper level, in our mental model. We are representing the state of a two-dimensional isotropic physical continuum by a set of variables situated in an orthogonal grid of the type presented in Fig.5.1 without much thought being devoted to the effect of this. It is known that the physical laws of conservation also hold in discretised form and that the differential form is just our way of making these laws more accessible to our own mental processes, so that from this point of view there is no problem with discretisation as such. However, this is true only so long as the connectivity is not affected, and what we usually achieve with the two-dimensional discretisation is to change the connectedness of the points in the domain. As its result, the points along the lines parallel with the principle axes preserve a strong connectedness while the points in any other directions maintain their connectedness only through the chains of other points which lie in the directions of the main axes. As long as the domain is not limited, this need not influence the results appreciably. However, the domains in which we are usually interested are not unlimited, and the effects of their limitations are felt to a greater or lesser extent depending on the solution methods employed, the characteristics of the modelled domain, the flow characteristics and the discretisation parameters.

The ADI algorithm, due to the splitting of the calculation of the two-dimensional flow into two series of one-dimensional problems, suffers from the consequences of reduced connectedness: the use of more than one algorithm that solves the full two-dimensional flow-pattern in one piece. On the other hand, all the beneficial features of the ADI algorithm make it still a favourable tool for the solution of flows that are two-dimensional in plan. Thus an extension of the ADI algorithm has been developed, named ADI+, which addresses this problem and tries to improve the connectednness of the points of the discretised domain through introducing an adaptive behaviour.

5.4. The flow adaptive algorithm - ADI+

The laws of conservation of momentum as written above are invariant with respect to the choice of directions. This means that they are not strictly connected with the directions x_1 and x_2 but they hold invariantly for any other pair of orthogonal directions. These directions can as well be the directions x_3 and x_4 making angles of 45° and 135° with the direction x_1. If we now draw these directions on our original grid (see Fig.5.4) we see that these new lines connect as well the h-points of our original grid. The points along diagonals which were not directly connected in the original ADI grid schematised in Fig.5.1 now become connected. This changes the whole connectivity pattern of the discretisation grid and enables the construction of a much more robust finite difference schemes for the solution of two-dimensional flows.

We should emphasise that this is only one possibility out of an apparently endless set of possibilities for improving the connectivity of a calculation even in only two directions. We thus consider the scheme associated with the grid of Fig.5.4 as just one example of an adaptive scheme within the context of two-dimensional nearly-horizontal flow. We by no means exclude the possibility (and indeed consider it highly probable) that more efficient ways of achieving the same objective may be found- or possibly already exist without our being aware of them!

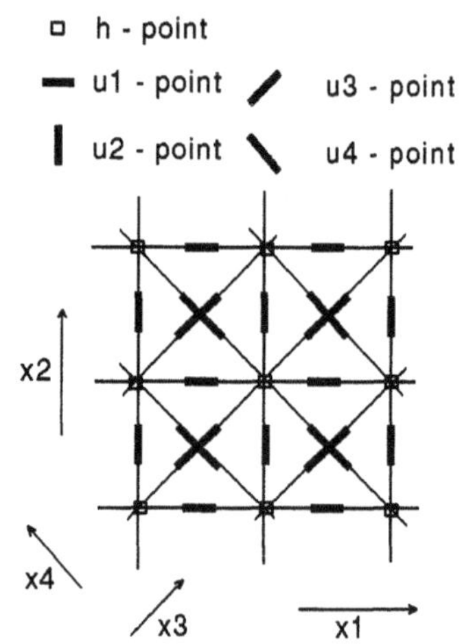

Fig.5.4

Clearly, then, two dimensional flow can as well be described by the shallow water equations in the new grid x_3-x_4:

$$\frac{\partial h}{\partial t} + \bar{h}\left(\frac{\partial u_3}{\partial x_3} + \frac{\partial u_4}{\partial x_4}\right) = 0 \tag{5.5}$$

$$\frac{\partial u_3}{\partial t} + g\frac{\partial h}{\partial x_3} = 0 \tag{5.6}$$

$$\frac{\partial u_4}{\partial t} + g\frac{\partial h}{\partial x_4} = 0 \tag{5.7}$$

with:

x_3, x_4 - space coordinates
u_3 - velocity in x_3 direction
u_4 - velocity in x_4 direction

If the dominant flow direction coincides with the direction x_3 or x_4, the solution based on the discretisation of equations (5.5) to (5.7) would clearly be better than the one obtained from the equations (5.2) to (5.4) in the original coordinate system defined by axes x_1 and x_2. Accordingly, an algorithm has been developed which can, on the basis of the flow direction, alternate between these two descriptions of flow, the one in the x_1-x_2 coordinate system and the other in the x_3-x_4 coordinate system. In order to make this alternation reasonably smooth, a form of weighted combination of two (and, as will later in fact appear, three) solutions was developed.

Based on these considerations, the flow adaptive algorithm ADI+ was developed as a combination of two ADI algorithms. Each of these is performed in a different grid direction and over different portions of the time step. The grids are unified through their h-points and the time steps are weighted portions of the original time steps. Each ADI algorithm splits its time step in two parts; thus, the ADI+ algorithm consists of four series (and indeed, in a sense that we shall introduce shortly, *six* series) of 'one-dimensional like' problems.

In the flow adaptive ADI+ algorithm, the description of the flow field is, then, divided between two coordinate systems. In each of these systems, it is further decomposed into components parallel to the coordinate axis. The total flow velocity is then seen as a vector sum of these four components. (See Fig.5.5 for an illustration). In the case when the flow coincides with any of the four directions, there should be a non-zero component only in that direction and all the others should vanish. In order to achieve this, the division between the two coordinate systems should itself be a function of the flow direction. Therefore a *weighting coefficient* which governs the division is defined as a function of the flow direction. This weighting coefficient will be more closely determined shortly.

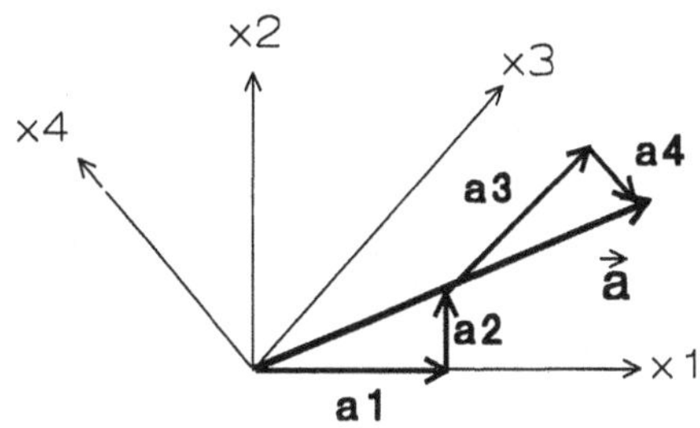

Fig.5.5

Let us first describe the algorithm. First the weighting coefficient (w) is calculated based on the flow field from the previous time step or the initial data. Then the flow velocity ($\vec{u}$) is divided into two parts $w\vec{u}$ and $(1-w)\vec{u}$ so that their sum is equal to the original velocity. Each of these vectors corresponds to one of the two coordinate systems, x_1-x_2 or x_3-x_4. Each of these two vectors is then decomposed into components in the directions of the corresponding coordinate axes. The vector $w\vec{u}$ is decomposed in the directions x_1 and x_2 into component u_1 and u_2 while the vector $(1-w)\vec{u}$ is decomposed into components u_3 and u_4 in directions x_3 and x_4 respectively.

Then the time step Δt is divided into two parts $w\Delta t$ and $(1-w)\Delta t$. The discretisation of the equations (5.2) to (5.4) is done on the $w\Delta t$ part of the time step while the discretisation of the equations (5.5) to (5.7) is done on the rest of the time step, $(1-w)\Delta t$. The details of the finite difference approximations are presented in appendix 5.2.

A schematised development of each variable throughout one time step is represented in Fig.5.6. In each of the four directions, two sets of unknowns are solved, these being the velocity in that direction and the water depth which is being accumulated. This feature is an extension of the classical ADI algorithm where in each of the directions - there are then only two - a flow velocity and a part of the water depth variation are determined. All the details about the ADI+ algorithm from the discretisation of the governing equations to the pseudo-code of the solution algorithm are described in appendix 5.2.

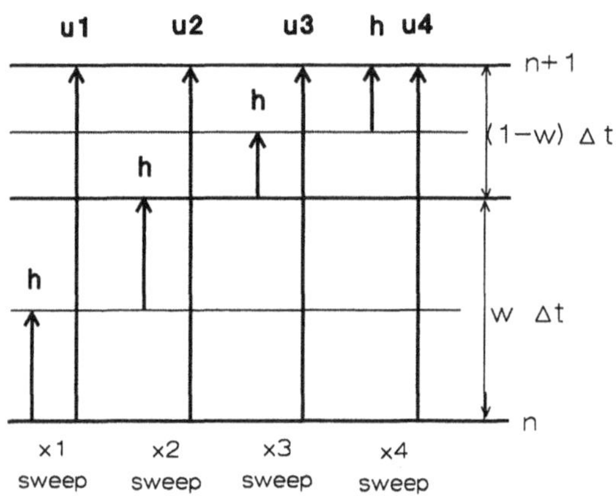

Fig.5.6

5.5 Feedback through the weighting coefficient

The weighting coefficient should of course depend on the flow direction in such a way that when the flow is in the x_1 or x_2 direction the full weight will be given to the solution obtained in the x_1-x_2 grid. On the other hand, when the flow entirely is in the x_3 or x_4 direction, the full weight should be given to the solution in the x_3-x_4 grid and no weight at all should be given to the solution in the x_1-x_2 grid. The weighting function then plays the role of a feed-back function which influences directly the core of the computation. The flow field is the input data for the weighting function. One possible weighting functions (albeit not computationally efficient!) is:

$$w = 0.25(2 + \cos 4\alpha - \cos 4\beta) ,$$

with:

$$\alpha = \arctan\left(\frac{u_2}{u_1}\right), \qquad \beta = \arctan\left(\frac{u_4}{u_3}\right)$$

which is represented graphically in Fig.5.7 for the case when $\beta = \alpha - 45°$.

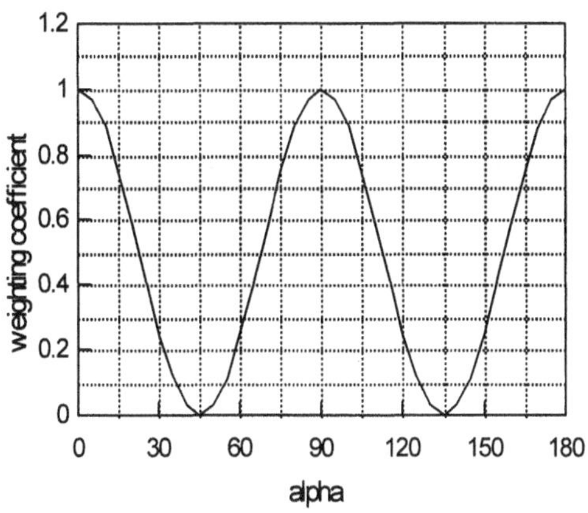

Fig.5.7

Although aware of its inefficiency, we have used this feedback function in the models developed for this study mainly due to its smoothness. As these models are just prototypes used to develop the new algorithm this is not seen as a major disadvantage. Moreover, it seems that the form of the weighting function has a considerable influence on the overall performance of the ADI+ algorithm. Thus, if a serious model is to be build based on the ADI+ algorithm, considerable attention should be paid to the development of an efficient and still smooth feedback function and a thorough investigation should be made into its influence on the overall algorithm.

5.6 The adaptive control structure

The adaptivity of this algorithm is based on the balance between solutions of the flow in the two pairs of directions, x_1-x_2 and x_3-x_4. The feedback function is evaluated on the base of the flow field results obtained at the previous time step as explained earlier. Thus, the whole system can be regarded as a closed loop system and represented by the schema shown in Fig.5.8. The adaptation transducer is a numerical operator which provides a weighting coefficient as the essential input to the numerical operator: this weighting coefficient is a typical example of an algebraic feedback function. The adaptation is primarily linked to the direction of flow, so that if there is no change in the flow direction, the NO will not be affected by the result obtained from the adaptation transducer.

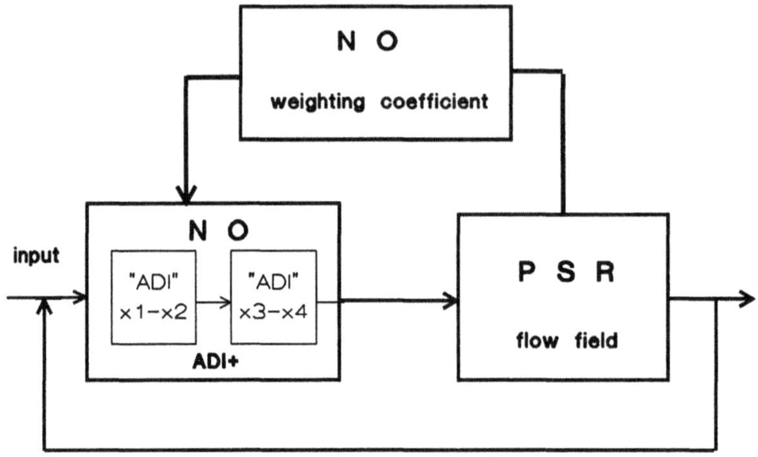

Fig.5.8

The PSR of the system is the set of all flow variables, which in this case are the water levels and velocities in the four different directions. The numerical transducer consists of two parts. Each of these parts is itself a numerical transducer and the weighting coefficient supplied by the adaptation transducer balances the application of these two numerical transducers. The way in which the numerical operator is build-up was already discussed, while all computational details are relegated to Appendix 5.2.

5.7 Some features of the ADI+ grid

Since the same *h*-points are involved in both grids (see Fig.5.4) and moreover these are involved in all four directions, the value of the water depth in *h*-points is being built-up out of four contributions within one time step. On the other hand, values of velocities in each of the four directions are obtained just from the solutions in the corresponding directions. The type of grid that is used here and which is schematised in Fig.5.4 is said to accumulate mass at common points. This is a method that is commonly introduced to prevent the formation of double solution structures (Abbott and Basco, 1989, p.268).

However, if only the grid in directions x_3-x_4 is considered it becomes clear that this grid is not of the same type as the original grid in the x_1-x_2 directions. As schematised in Fig.5.9, it is obvious that this grid is prone to generate double solutions, being in fact a composition of two partially, or even completely disconnected, grids. The most severe effects arising from this will in fact be experienced when the flow is entirely at 45°(or 135°) to the original grid, as in that case the original grid, which otherwise provides some connection between the concerned h points, is not used at all.

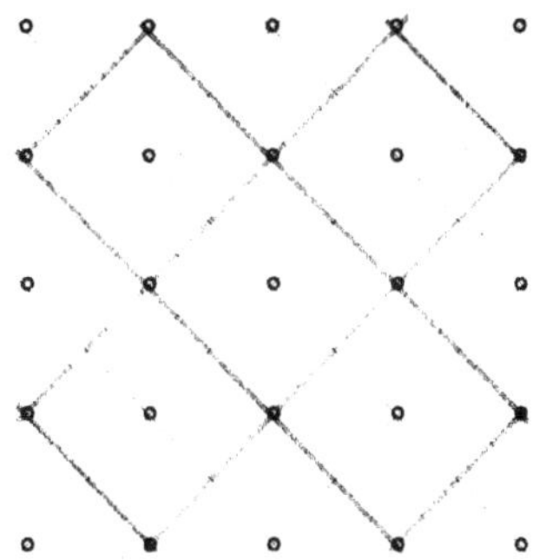

Fig.5.9

In order to avoid the development of this kind of double solution, another kind of weighted average interface can be applied to the results of the calculation after each time step. In this way all the initial oscillations which might develop into a double solution are 'washed out'. However, one must be aware that this averaging constitutes a sort of dissipative interface and therefore the results obtained may demonstrate some dissipation as well.

The simplest form of averaging, and the one employed when following this approach in the present study, can be represented by the formula:

$$a_{j,k} = (1-\alpha)\, a_{j,k} + \alpha\, a_{j-1,k} + \alpha\, a_{j+1,k} + \alpha\, a_{j,k-1} + \alpha\, a_{j,k+1} \;, \qquad 0 \le \alpha \le 1/4 \qquad (5.8)$$

The other possibility for avoiding the formation of a double solution structure is the introduction of additional h points at the intersections of the grid lines in the x_3 and x_4 directions. A grid of this kind is schematised in Fig.5.10. The grid is then transformed into the same form as the original grid and the danger of generating double solutions is diminished.

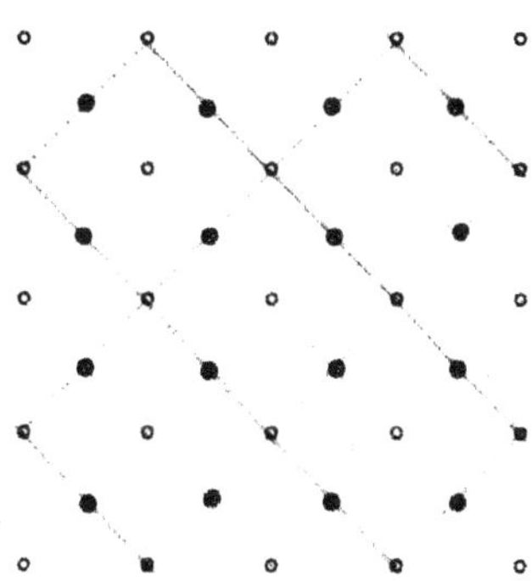

Fig.5.10

This approach has disadvantages as well. The values of water depth in these additional *h*-points are not being built up through all four directions of computation but only along two of these, namely those of x_3 and x_4. This may also lead to double solutions. Again a sort of averaging can be applied in order to approximate the values of the water depth at these points at the beginning of the $(1-w)\Delta t$ part of the time step. This time the value of the considered variable at the end of the previous time step is corrected for the average of the change of variables in the four neighbouring points within the $w\Delta t$ part of the time step. If there were no change of these variables within this period, as for example in the steady state or in the case when the weighting coefficient was equal to zero, then no correction would be applied.

The introduction of these additional grid points also allows for a much more precise positioning of the boundary conditions in the case of boundaries laid down at 45° or 135° to the x_1 axis. This is a feature that one would in fact expect from the ADI+ algorithm in any case, but the fact that the x_3-x_4 grid is not of the same type as the standard grid originally prevented any improvement, and it even acted in the negative way.

In order to illustrate the nature of the improvement, the three schemes are presented in Fig.5.11 to Fig.5.13. In each of these a closed boundary condition along the line laid at an angle of 45° to the x_1 axis is sketched. This is done in Fig 5.11 for the standard ADI algorithm, and this applies also to the x_1 and x_2 direction of any ADI+ algorithm. The position of the boundary conditions in the x_3 and x_4 directions of the ADI+ algorithm without additional *h* points is schematised in Fig 5.12 while in Fig.5.13 the same is presented for the ADI+ algorithm with additional h points. The other boundary condition types can be positioned in a similar way.

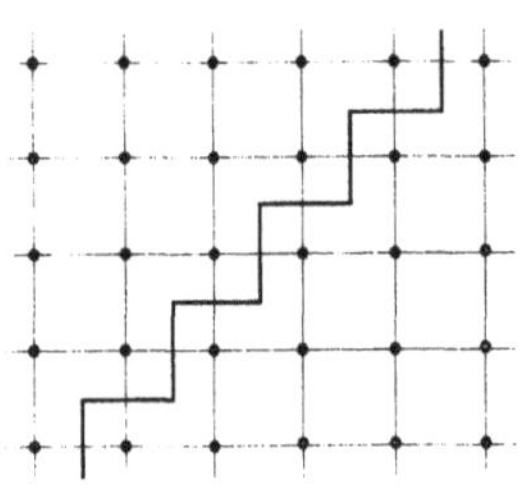

Fig.5.11

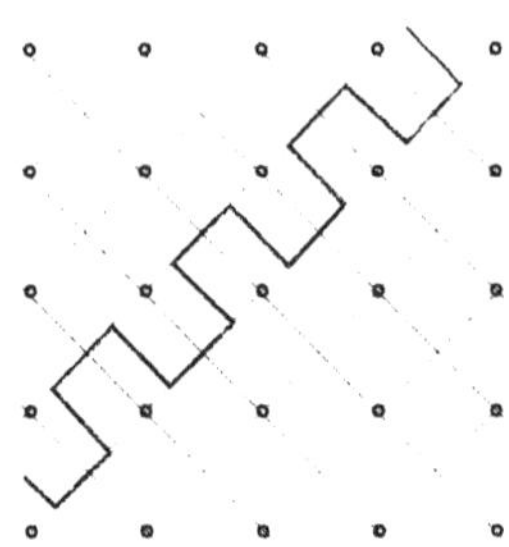

Fig.5.12

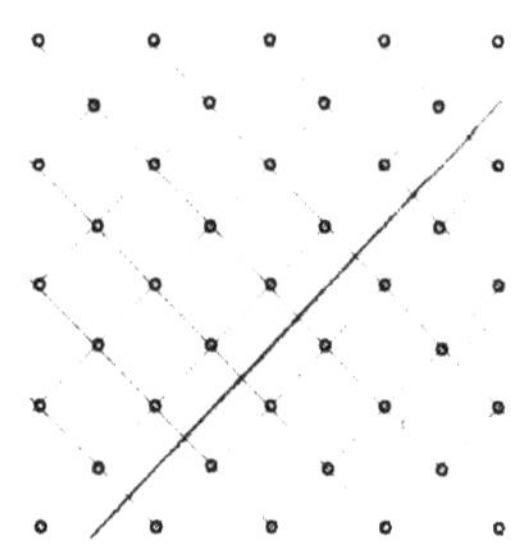

Fig.5.13

From the above figures it is clear that the use of additional h points improves considerably the definition of the boundary conditions, which has long being recognised as one of the principle sources of errors implicit in the use of the ADI algorithm (Weare, 1986).

Based on these considerations, four types of models were developed, each following one of these different approaches.

5.8 Models and tests

Three different models were developed based on the ADI+ algorithm described above and whose computational details are given in Appendix.5.2. Besides these, a model was also developed based on the standard ADI algorithm. When used for comparison, it is simply called **ADI**.

The first model is one that is simply based on the ADI+ algorithm without any modifications directed to reducing the oscillations caused by the grid type. This model is called **ADI+1**.

The second model includes an averaging interface introduced in order to avoid spurious oscillations arising from the double solution structure in the x_3 and x_4 grids. In its first version the averaging proceeding according to equation (5.8) was applied only to the h points. When this appeared to be insufficient, the same form of averaging was applied also to the velocities u_3 and u_4. It is the results of this later version, called **ADI+A,** that are presented in this chapter.

The third model, called **ADI+H**, has additional h points introduced at the intersections of the grid lines in the x_3 and x_4 directions. As a consequence of the introduction of theses additional h points, the boundaries laid at an angle of 45° to the main axis are defined according to the sketch shown in Fig.5.13. In the other two models, this is introduced according to the sketch of Fig.5.12.

Various steady and unsteady tests were performed with these models during the debugging and investigation phases. Here, only the results of two representative tests will be given although later in this chapter, a series of tests with various narrow channels is presented. The tests given here are the two unsteady flow tests situated in the two different domains, both with zero-flow initial conditions and with velocities given as a function of time as the open boundary conditions.

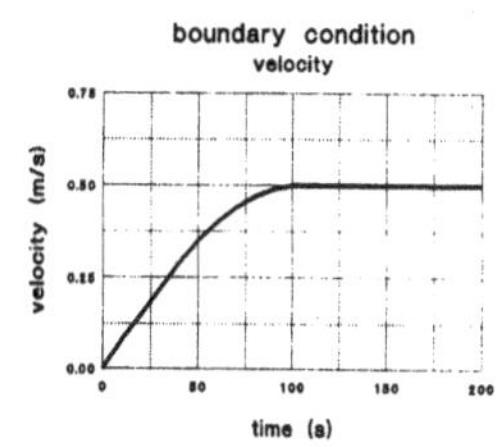

Fig.5.14

The velocities as functions of time are shown in Fig.5.14. The two representative domains are the straight channel, laid down at 45° to the x_1 axis, as presented in Fig.5.15, and the L shaped channel presented in Fig. 5.16.

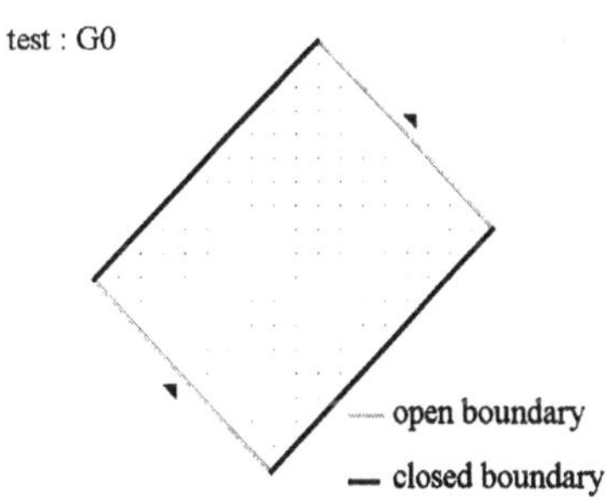

Fig.5.15

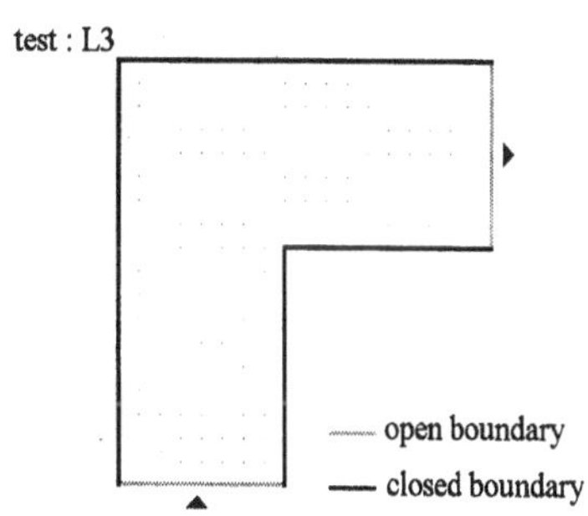

Fig.5.16

The test G0 was simulated for 500 seconds (50 time steps) with four different models and the final flow fields of these simulations are presented in Fig.5.17 to Fig.5.20.

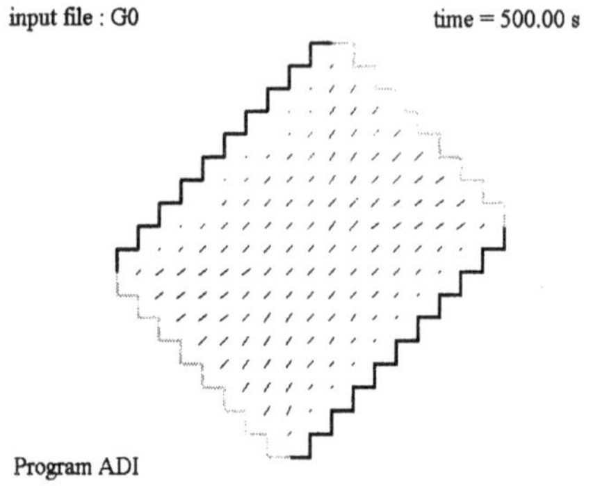

Fig.5.17 Program ADI ,test G0

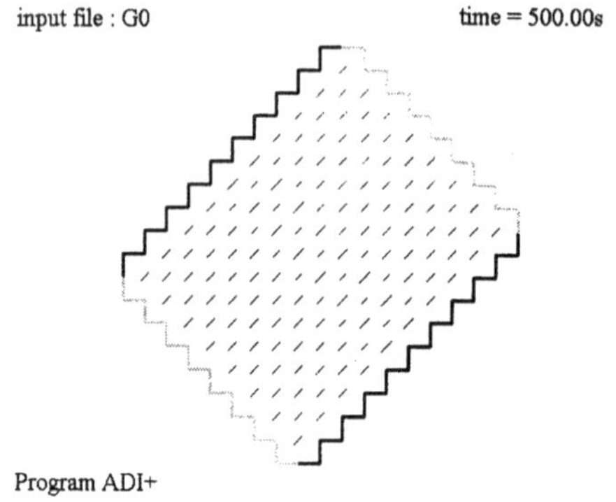

Fig.5.18 Program ADI+, test G0

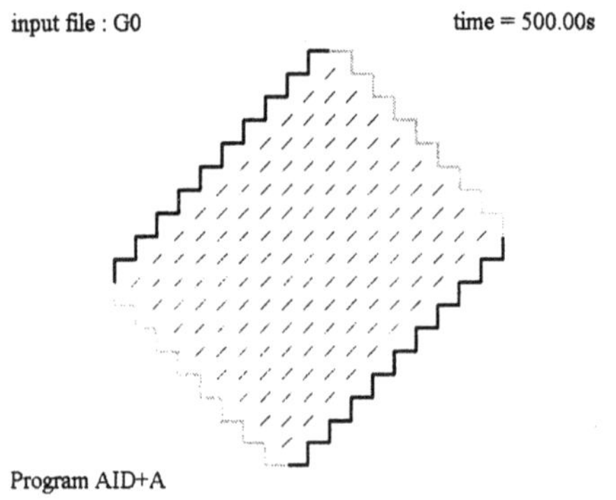

Fig.5.19 Program ADI+A, test G0

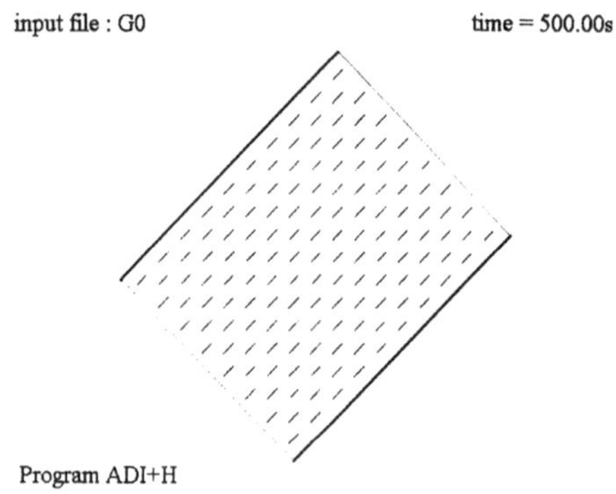

Fig.5.20 Program ADI+H, test G0

In the Fig.5.17 the flow field obtained by the standard ADI algorithm is presented. As can be seen, the flow field is symmetric but not completely uniform as a consequence of the unfavourable direction of the boundary conditions and flow with respect to the directions of the main axes. This result was, however, expected due to the problems associated with the standard ADI algorithm. In the Fig.5.18, where the flow field obtained by the model ADI+is shown, a clear double solution is observed. As discussed earlier, this is a consequence of the existence of two disconnected grids in the x_3-x_4 directions.

Fig.5.19 and Fig.5.20 representing results of simulations by the models ADI+A and ADI+H show practically no difference. The only different feature is that the model ADI+H allows a description of boundary conditions which is more suitable for this test. This resulted in the reduction of oscillations originating from irregular boundaries but these were obvious only in the clear unsteady period of simulation (time steps 0 till 10) and by the 50th time step they were already washed out.

In Fig.5.21 to Fig.5.22 the flow field results of test L3 as obtained by the different models are presented. All the results are given for the time of 400 seconds (40 time steps) because the model ADI+, whose results are given in Fig.5.22, goes unstable if the simulation time is extended. This instability is a consequence of the double solutions originating from the two disconnected grids in the x_3-x_4 directions. When the two grids are connected, as in models ADI+A and ADI+H, the algorithm is stabilised.

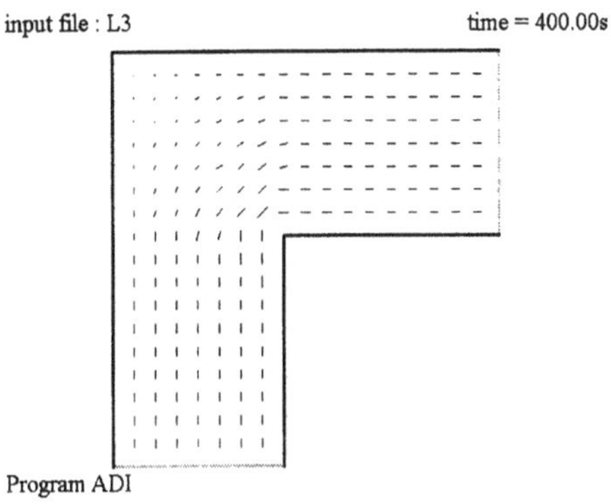

Fig.5.21 Program ADI, test L3

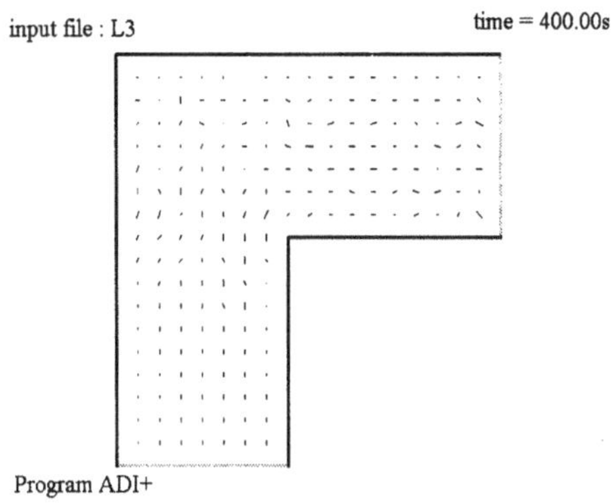

Fig.5.22 Program ADI+, test L3

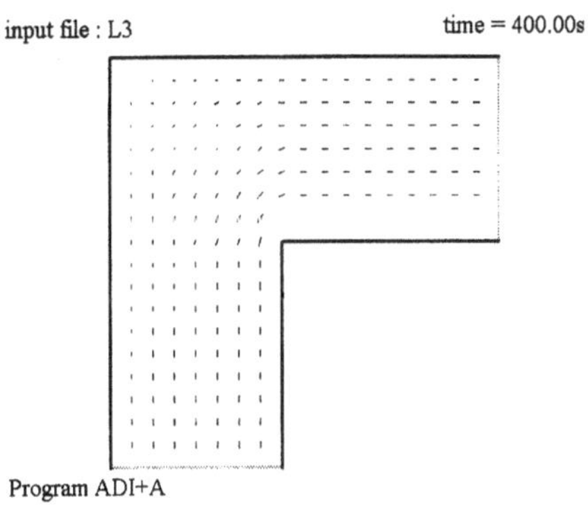

Fig.5.23 Program ADI+A, test L3

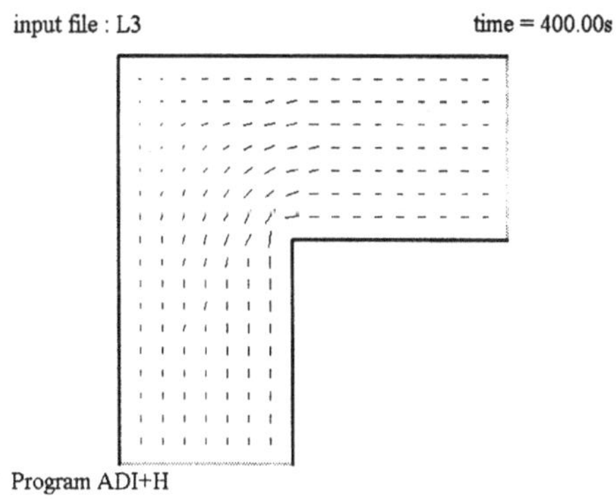

Fig.5.24 Program ADI+H, test L3

The results presented in Fig.5.21 and the those presented in Fig.5.23 and Fig.5.24 show some differences mainly in the flow pattern in the vicinity of the inner corner. However, in order to provide any definitive conclusions about predictive capability of those models and consequently about the preferential choice between the models ADI+A and ADI+H, more extensive testing will need to be performed.

5.9 Narrow channels

The following example was set up in order to test the performance of the ADI+ algorithm in the case of a narrow channel laid down at the angle of 45° to the x_1 axis. There are in fact four different domains that were tested, each of them with a different width of the channel. The range of widths was $2\Delta x$, $4\Delta x$, $6\Delta x$ and $8\Delta x$ in the x_1 direction. Two of them, with widths $2\Delta x$ and $8\Delta x$, are represented in Fig.5.25 and Fig.5.26 respectively. The length of channel is in all the examples equal to 20 grid points along the diagonal.

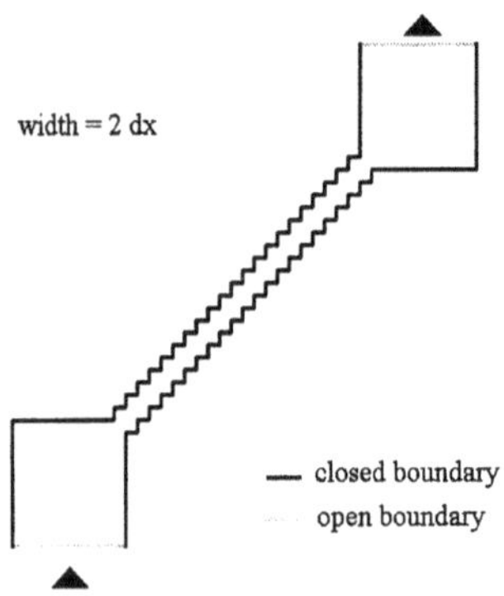

Fig.5.25

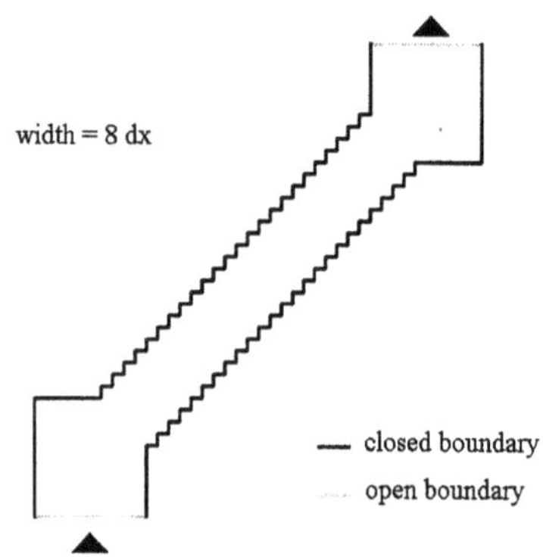

Fig.5.26

The initial conditions for all the tests were a zero velocity in all the directions and a constant water depth of 2m all across the domain. The size of the space step is 10m. At the open boundary conditions, water depth is given as a function of time. At the outgoing boundary it is a constant water depth equal to the initial conditions, while at the incoming boundary it is a variable according to the graph shown in Fig.5.27.

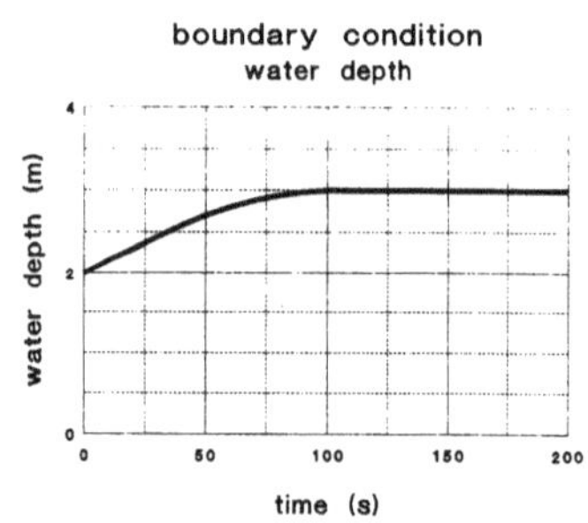

Fig.5.27

Results

First the model based on the classical ADI algorithm was applied to these tests. Along the channel with the width equal to two space steps (Fig.5.25) there was no spreading of the wave. The whole channel and the upper pool stayed at the initial water depth of 2 metres while the lower pool filled up to the depth of 3 metres. The other three tests have not shown such drastic features, but it was observed that the time needed for the wave to reach the last point along the channel does vary with the size of the time step. In Fig.5.28 the relationships between the time step and the time needed for the initial disturbance to reach the last point along the channel are shown for different widths of the channel. If the same relationships are expressed in terms of the number of time steps needed, as given in Fig.5.29, than it can be observed that for each width of the channel there is a minimal number of time steps required for the disturbance to travel. This means that if larger time steps are employed, the results are less realistic.

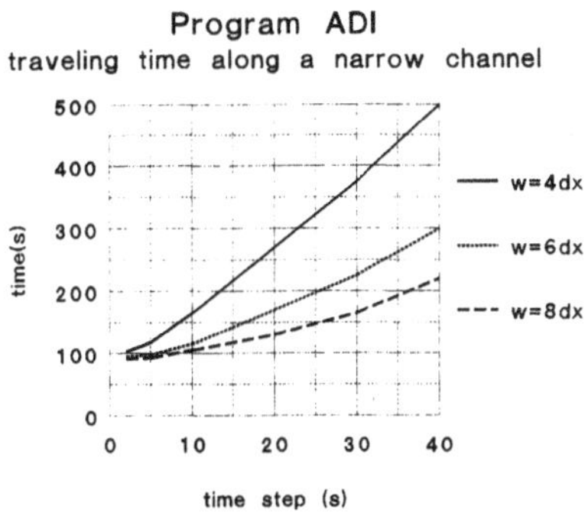

Fig.5.28

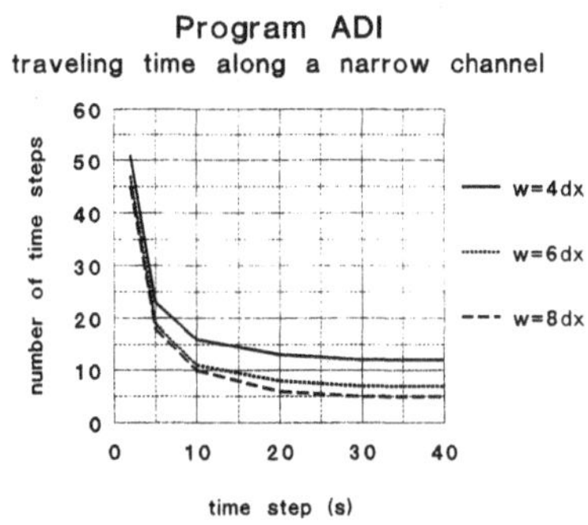

Fig.5.29

However, if a program based on the ADI+ algorithm is applied to these tests (in this case it was program ADI+A) the time needed for the initial disturbance to reach the other and of the channel is approximately 100s for all four widths of the channel including the case with a channel width of only $2\Delta x$. This result is independent of the size of the time step. This feature of the ADI+ algorithm seems to allow the use of much larger time steps then does the classical ADI algorithm.

In the following four figures (Fig.5.30 till Fig.5.33), longitudinal profiles along the axis of the channel are shown. In each of the figures, results of the ADI program are compared with those obtained by using the ADI+A program. All these tests were performed with equal time steps, Δt=20s, and the results correspond to an elapsed time of 100s, which thus corresponds to five time steps.

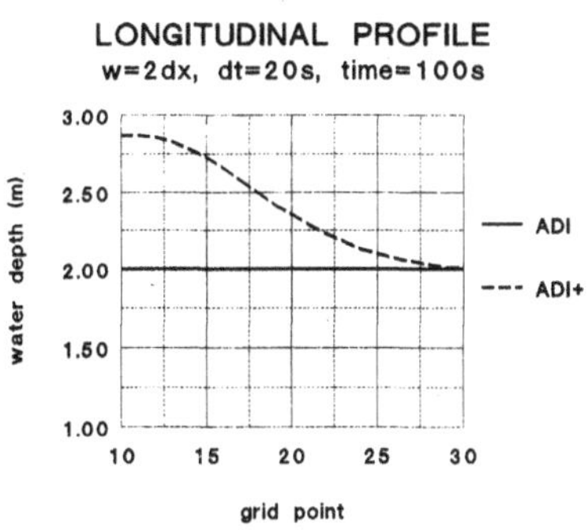

Fig.5.30

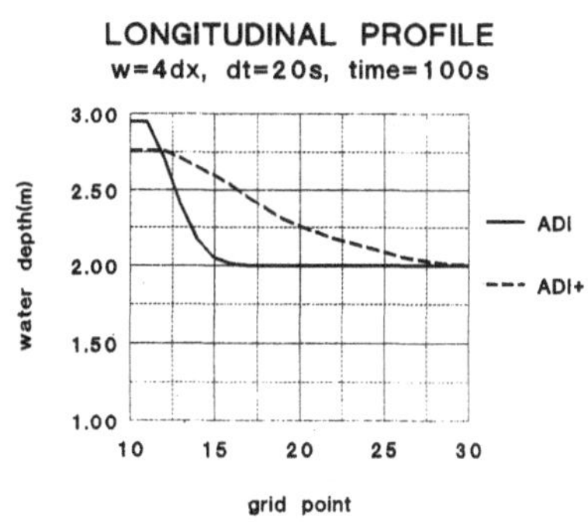

Fig.5.31

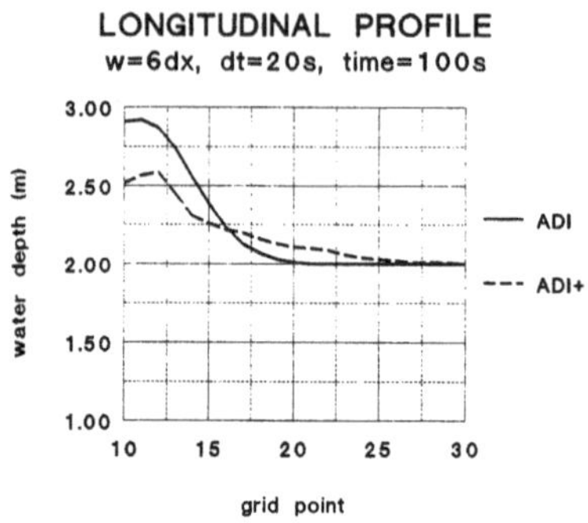

Fig.5.32

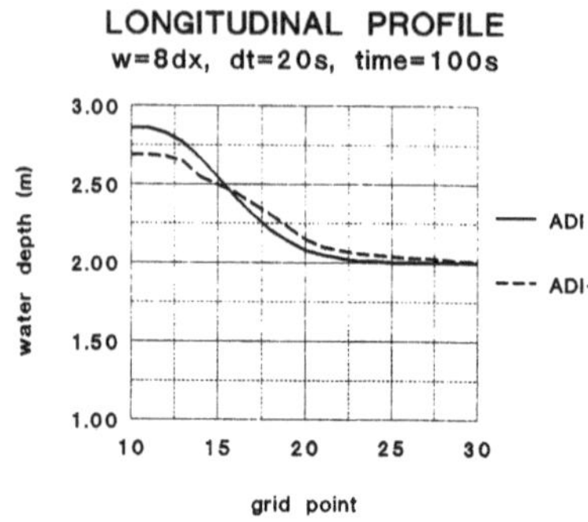

Fig.5.33

From these figures it can clearly be observed how the standard ADI algorithm hampers the spreading of the wave along the narrow channel. It can also be seen that this effect is proportional to the narrowness of the channel, where in the limiting case (Fig.5.30) of the channel with width of $2\Delta x$ there is no flow in the channel at all. For the ADI+ algorithm, the narrowness of the channel at this orientation does not represent any problem.

5.10 Conclusions

Although there are still many open questions in its area of applications, the models presented in this work based on the ADI+ algorithm do open the door to the development of new forms of adaptive algorithms for the solution of flows that are two dimensional in plan.

At present, the clear shortcomings of the algorithm are the form of the 'rotated' grid (the grid in the x_3-x_4 directions) and the feedback function used in the adaptation feedback. The rotated grid has a form which differs from the standard form of the ADI grid and if no special treatment is included this leads to instabilities in unsteady flow simulations. A preferable way to approach this problem would now appear to be, not to start from the standard ADI grid and then introduce the rotated grid, but to start the building up of both grids simultaneously, bearing in mind from the beginning the kind of algorithm for which the grid is going to be used. The viability of introducing the feedback function (the weighting coefficient) depends primarily on its complexity, but it might as well be questioned whether an algebraic function is the best option, or whether a logical operator might serve better instead.

Although none of the solutions introduced here seems ideal, the one with additional h points seems preferential. Firstly it requires less averaging for the stabilisation of the results, and besides that it offers a great improvement in the treatment of the boundary conditions for boundaries laid down at an angle of 45° to the main axes.

As the ADI+ algorithm requires a considerably heavier computational effort than does the standard ADI algorithm, it should perform at least at the same level of accuracy with two times larger time steps, which at this stage of its development still does not seem fully attainable. However, in the tests with narrow channels, the ADI+ algorithm showed its very good characteristics and in such cases it seems feasible to use significantly larger time steps with the ADI+ algorithm that with the ADI algorithm. Hence, the way forward might be to limit the application of this adaptive algorithm to regions for which it is particularly suited, such as narrow channels laid at 45° degrees, or to regions with exigent boundaries or regions with very variable flow patterns.

A considerable gain in performance might be achieved if the feedback function were to be chosen in such a way as to act more as a kind of switch between the calculations in the two grids, namely the x_1-x_2 grid and the x_3-x_4 grid. This could be achieved with the weighting coefficient being set to one or zero over most of the flow field, with intermediate values being computed over short intervals in order to enhance the smoothness. This would result in the reduction of the calculation effort since only in a limited number of cases would the calculation in both grids be performed, and in the most of the cases only the calculation in the better-suited grid would be sufficient.

In short, the flow adaptive algorithm ADI+, although still at a very early stage in its potential development, shows promising capabilities for the introduction of adaptive algorithms into the simulation of two-dimensional nearly-horizontal flows.

CHAPTER 6

General Conclusions

Computational hydraulics continues to pose its problem of describing what we observe in our outer world of fluid flow in terms that our digital machine can best comprehend. Hydroinformatics now however places this problem within the much wider context of how we can best apply computational hydraulics, together with all our other technologies, to improving the balance between the needs of our own, human economies and the needs of nature. By positioning computational hydraulics in this way, hydroinformatics is able to place ever heavier demands upon computational hydraulics simply because it provides the physical means to develop even more refined and advanced methods for meeting these increasing demands.

Flow adaptive schemes can be viewed within the perspective of this ongoing development. The increasing demands of hydroinformatics make the use of modelling systems, such as we see them in current engineering practice, less and less acceptable. Eventually the fifth generation of such systems, integrable in larger informatics structures, will emerge. Adaptive schemes are one of first, but important steps, in this direction. The examples that have been used to open up this field of investigation have all arisen from the expanding needs of practice, while the added complication that flow adaptation introduces as it response to these needs is justified, both in economic and in more general social terms, by the increased range of applications that hydroinfomatics thereby provides.

Each of the four cases investigated in this work is itself a development in computational hydraulics. These developments were motivated by real shortcomings of currently used methods and they have all, albeit to different extent, provided some improvements. It is only when removed from the immediate context of each of these problems, however, that one can realise their common ground. In each of the cases a kind of numerical scheme is developed which uses intermediate flow results, or at least the initial data, in order to adapt itself, and by so doing achieve a more efficient performance. This property, here termed 'flow adaptivity', seems to be a promising trend in future developments of computational hydraulics, as this work indicates.

In each of the cases studied the scheme becomes more flexible, more 'parameterised' with the flow variables, so as to suit itself better to more complicated physical situations. It is in this sense that the scheme becomes deconstructed, or 'deconstructs itself'. The present study provides a certain kind of exploration of the processes that occur during such a deconstruction of numerical schemes for flow problems.

The main innovative conclusion of the presented work is to realise the principle of adaptivity, to formulate it and make clear that its application will in future become general and not any more a special or particular approach. This is only a very partial exploration because of the immense range of possible kinds of flow adaptive schemes that can arise in so many different fields of applications. Indeed, when reviewing the immensity of this range, it must appear as if the present work is really only 'scratching the surface' of the subject. This effort may however still serve to indicate the possibilities inherent in flow-adaptive schemes and some of the means that are available for realising them in practice.

APPENDIX 2.1

Algorithm for the determination of velocity profiles in flows with additional resistance due to vegetation

The numerical model for the determination of the velocity profile in the flow with additional resistance is based on the equation of conservation of momentum in the horizontal direction, which for the present purpose can be most conveniently written as a differential-difference equation:

$$\frac{\partial u}{\partial t} + u\frac{\partial u}{\partial x} + w\frac{\partial u}{\partial z} + \frac{1}{\rho}\frac{\partial P}{\partial x} - gi - \frac{1}{\rho}\frac{\partial \tau}{\partial z} + \frac{F_x}{\rho \Delta x \Delta z} = 0 \qquad \text{(A.2.1.1)}$$

with:

u	- horizontal velocity	i	- bottom slope
w	- vertical velocity	τ	- shear stress
P	- pressure	$\frac{F_x}{\rho \Delta x \Delta z}$	- additional horizontal drag forces
ρ	- density of water		
g	- gravity acceleration		

Since the model is developed in order to study the effects of additional resistance in steady uniform flow, some of the terms from equation (A.2.1.1) are neglected. These are the pressure term and the convective momentum term. The reduced equation then has the form:

$$\frac{\partial u}{\partial t} - gi - \frac{1}{\rho}\frac{\partial \tau}{\partial z} + \frac{F_x}{\rho \Delta x \Delta z} = 0 \qquad \text{(A.2.1.2)}$$

The numerical scheme used for solving this equation is an implicit finite difference scheme. This maintains the stability of the scheme regardlessly of the choice of such discretisation parameters as the time and the space step. The model is one-dimensional in the vertical direction following the discretisation grid shown in Fig.A.2.1.1 with *kk* being the number of grid points. The unknown velocities (u_k) are placed in all the discretisation points.

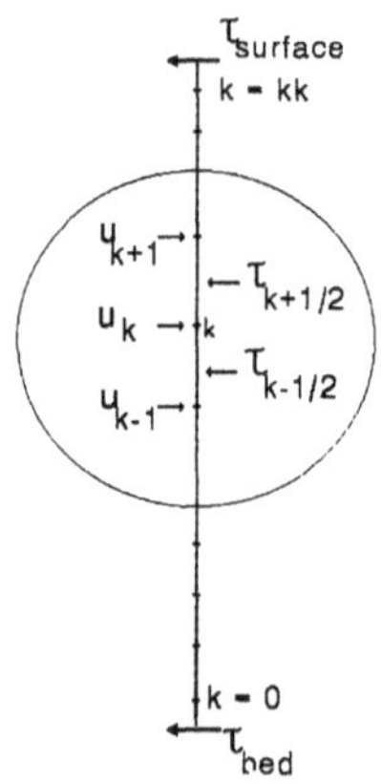

Fig.A.2.1.1
the discretisation grid

Due to the two different descriptions of the shear stress which are used in the model and which will be explained later it is important to maintain a discretisation point at the boundary between these two regions. The height of the lower region is a linear function of the reed height (h_r) whose coefficient *p* is to be determined by calibration of the model. Therefore, the grid is divided into two parts, the first one, from the bottom up to the height ph_r and the second one from that point up to the water surface. This division is schematised in Fig.A.2.1.2.

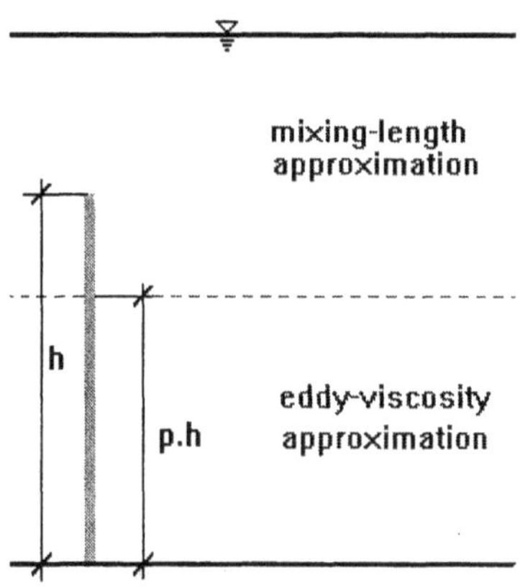

Fig.A.2.1.2

In each of these parts separately the space step is constant, but these two space steps usually differ. If k_1 and k_2 are the numbers of discretisation points in each of the parts, then the grid steps are defined by :

$$\Delta z_1 = \frac{p \cdot h_r}{k_1 + 0.5} \qquad \text{and} \qquad \Delta z_2 = \frac{h_w - p \cdot h_r}{k_2 + 0.5}$$

The acceleration term

Although the model is applied to constructing solutions for steady state conditions, it is in principle an unsteady model. This feature makes possible the calculation of the unknown steady-state condition starting from any initial conditions, even though these may be unrealistic, on the basis that these initial influence will be 'washed away', so to say, in time. We write simply:

$$\frac{\partial u}{\partial t} \sim \frac{u_k^{n+1} - u_k^n}{\Delta t}$$

The shear stress term

In the layer of water above the point ph_r a turbulent shear stress approximated by a mixing length theory is employed, so that.

$$\tau(z) = \rho\, l^2 \frac{\partial u}{\partial z} \left| \frac{\partial u}{\partial z} \right|$$

where l is the mixing length at the level z determined by:

$$l = \kappa z \left(1 - \frac{z}{h} \right)^{0.5}$$

with κ being the von Karman coefficient and h being the water depth.

The shear stress in the lower layer is modelled according to eddy-viscosity theory as:

$$\tau(z) = \epsilon \frac{\partial u}{\partial z}$$

where

$$\epsilon = \rho \cdot \alpha \cdot s \cdot u$$

with α being an empirical coefficient and s being the distance between the reeds.

The shear stress term is then discretised in three different ways dependant on the relative position of the discretisation point within these two regions.

$0 < k < k_l$ - inside the lower layer

$$\frac{1}{\rho}\frac{\partial \tau}{\partial z} \approx \frac{\rho \alpha s}{\rho \Delta z}\left(\frac{u_{k+1}^{n} + u_{k}^{n}}{2} \frac{u_{k+1}^{n+1} - u_{k}^{n+1}}{\Delta z} - \frac{u_{k}^{n} + u_{k-1}^{n}}{2} \frac{u_{k}^{n+1} - u_{k-1}^{n+1}}{\Delta z} \right)$$

$k = k_l$

$$\frac{1}{\rho}\frac{\partial \tau}{\partial z} \approx \frac{1}{\rho \Delta z}\left(\rho l^2_{k+\frac{1}{2}} \left| \frac{u_{k+1}^{n} - u_{k}^{n}}{\Delta z} \right| \frac{u_{k+1}^{n+1} - u_{k}^{n+1}}{\Delta z} - \rho \alpha s \frac{u_{k}^{n} + u_{k-1}^{n}}{2} \frac{u_{k}^{n+1} - u_{k-1}^{n+1}}{\Delta z} \right)$$

$k_l < k < kk$ - inside the upper level

$$\frac{1}{\rho}\frac{\partial \tau}{\partial z} \approx \frac{1}{\rho \Delta z}\left(\rho l^2_{k+\frac{1}{2}} \left| \frac{u_{k+1}^{n} - u_{k}^{n}}{\Delta z} \right| \frac{u_{k+1}^{n+1} - u_{k}^{n+1}}{\Delta z} - \rho l^2_{k-\frac{1}{2}} \left| \frac{u_{k}^{n} - u_{k-1}^{n}}{\Delta z} \right| \frac{u_{k}^{n+1} - u_{k-1}^{n+1}}{\Delta z} \right)$$

This feature clearly already introduces an element of adaptability into the model.

The additional force term

The additional force term in equation (A.2.1.2) is then approximated as follows:

$$F_k^{n+1} = \frac{m \cdot CD}{2} \cdot d_k \cdot hr_k^n \cdot u_k^n \cdot u_k^{n+1}$$

where: m - is the density of reeds per m^2,
CD - is the drag coefficient,
d_k - is the diameter of the reed at the level k ,
hr_k^n - is the portion of the reeds' effective height corresponding to the grid level k at the time level n,
u_k^n , u_k^{n+1} - are flow velocities at the level k at the respective time levels, n and $n+1$.

The boundary conditions

For this model, two boundary conditions are required, one at the bed and the other at the water surface. The positions of these two boundary conditions are marked on the grid shown in Fig.A.2.1.1.

The shear stress at the bed is given through the averaged velocity and the Chezy coefficient.

$$\tau_{-\frac{1}{2}} = g\frac{\bar{u}^2}{C^2}$$

At the water surface the slip boundary condition with no shear stress is introduced.

$$\tau_{kk+\frac{1}{2}} = 0$$

Both boundary conditions are introduced directly into finite difference approximations of the shear stress terms at the corresponding grid points i.e $k=0$ and $k=kk$.

The system of equations

These finite difference approximations are introduced in equation (A.2.1.2) at each discretisation point so that the resulting linear algebraic equation can be written as:

$$A_k u_{k-1}^{n+1} + B_k u_k^{n+1} + C_k u_{k+1}^{n+1} = D_k \qquad \text{(A.2.1.3)}$$

The expressions for coefficients A_k ,B_k , C_k and D_k now however differ according to the position of the point k with respect to the boundaries and the regions supporting different descriptions of the shear stress. Consequently, seven different sets of expressions are employed in this model.

For $k = 0$ and $kl = 0$:

$$A_0 = 0 \quad , \qquad B_0 = 1 + \frac{\Delta t}{\Delta z_2^{\,3}} \, l_{\frac{1}{2}}^2 \left(u_1^n - u_0^n \right) + CD \, d_0 \, hr_0^n \, m \frac{\Delta t}{2 \Delta z_2} u_0^n$$

$$C_0 = -\frac{\Delta t}{\Delta z_2^{\,3}} \, l_{\frac{1}{2}}^2 \left(u_1^n - u_0^n \right) , \qquad D_0 = u_0^n + i \, g \, \Delta t - g \frac{\Delta t}{\Delta z_2} \frac{\bar{u}^2}{C^2}$$

For $k = 0$ and $kl \neq 0$:

$$A_0 = 0 \quad , \qquad B_0 = 1 + \frac{\Delta t}{\Delta z_1^2} \alpha \, s \, \bar{u} + CD \, d_0 \, hr_0^n \, m \frac{\Delta t}{2 \Delta z_1} u_0^n$$

$$C_0 = -\frac{\Delta t}{\Delta z_1^{\,2}} \alpha \, s \, \bar{u} \; , \qquad D_0 = u_0^n + i \, g \, \Delta t - g \frac{\Delta t}{\Delta z_1} \frac{\bar{u}^2}{C^2}$$

For $0 < \underline{k} < kl$:

$$A_k = -\frac{\Delta t}{\Delta z_1^{\,2}} \alpha \, s \, \bar{u} \; , \qquad B_k = 1 + \frac{2 \, \Delta t}{\Delta z_1^2} \alpha \, s \, \bar{u} + CD \, d_k \, hr_k^n \, m \frac{\Delta t}{2 \Delta z_1} u_k^n$$

$$C_k = -\frac{\Delta t}{\Delta z_1^{\,2}} \alpha \, s \, \bar{u} \; , \qquad D_k = u_k^n + i \, g \, \Delta t$$

For $k = kl$:

$$A_k = -\frac{\Delta t}{\Delta z_1^2}\alpha s \bar{u} \ ,$$

$$B_k = 1 + \frac{\Delta t}{\Delta z_2^3} l^2_{k+\frac{1}{2}} \left| u^n_{k+1} - u^n_k \right| + \frac{\Delta t}{\Delta z_1^2}\alpha s \bar{u} + CD\, d_k hr^n_k m \frac{\Delta t}{\Delta z_1 + \Delta z_2} u^n_k \ ,$$

$$C_k = -\frac{\Delta t}{\Delta z_2^3} l^2_{k+\frac{1}{2}} \left| u^n_{k+1} - u^n_k \right| , \qquad D_k = u^n_k + i\, g\, \Delta t$$

For $kl < k < kk$:

$$A_k = -\frac{\Delta t}{\Delta z_2^3} l^2_{k-\frac{1}{2}} \left| u^n_k - u^n_{k-1} \right| ,$$

$$B_k = 1 + \frac{\Delta t}{\Delta z_2^3} l^2_{k+\frac{1}{2}} \left| u^n_{k+1} - u^n_k \right| + \frac{\Delta t}{\Delta z_2^3} l^2_{k-\frac{1}{2}} \left| u^n_k - u^n_{k-1} \right| + CD\, d_k hr^n_k m \frac{\Delta t}{2\Delta z_2} u^n_k \ ,$$

$$C_k = -\frac{\Delta t}{\Delta z_2^3} l^2_{k+\frac{1}{2}} \left| u^n_{k+1} - u^n_k \right| , \qquad D_k = u^n_k + i\, g\, \Delta t$$

For $k = kk$ and $kl < kk$:

$$A_{kk} = -\frac{\Delta t}{\Delta z_2^3} l^2_{kk-\frac{1}{2}} \left| u^n_{kk} - u^n_{kk-1} \right| ,$$

$$B_{kk} = 1 + \ + \frac{\Delta t}{\Delta z_2^3} l^2_{kk-\frac{1}{2}} \left| u^n_{kk} - u^n_{kk-1} \right| + CD\, d_{kk} hr^n_{kk} m \frac{\Delta t}{2\Delta z_2} u^n_{kk} \ ,$$

$$C_{kk} = 0 \ , \qquad\qquad D_{kk} = u_{kk}^{n} + i\, g\, \Delta t$$

For $k = kk$ and $k1 = kk$:

$$A_{kk} = -\frac{\Delta t}{\Delta z_1^{\,2}}\, \alpha\, s\, \bar{u} \ , \qquad B_k = 1 + \frac{\Delta t}{\Delta z_1^{2}}\, \alpha\, s\, \bar{u} + CD\, d_{kk}\, hr_{kk}^{\,n}\, m\, \frac{\Delta t}{2\, \Delta z_1}\, u_{kk}^{n} \ ,$$

$$C_{kk} = 0 \ , \qquad\qquad D_{kk} = u_{kk}^{n} + i\, g\, \Delta t$$

In effect, these seven sets of conditions superimpose an even more extended 'if - then - else' structure on the numerical schemes, so that it corresonds to a particularly extended control operator.

The system of equations containing kk equations of the type (A.2.1.3) has as unknowns the set of horizontal velocities u_k , $k = 0,..kk$. The matrix of this system is tri-diagonal and it is solved with a double-sweep algorithm consisting of an elimination and a substitution sweep. The recurrence relations for the elimination sweep are:

$$u_k^{n+1} = E_k \cdot u_{k+1}^{n+1} + F_k \tag{A.2.1.4}$$

The initial coefficients E_k and F_k are given by:

$$E_1 = -\frac{C_1}{B_1} \qquad \text{and} \qquad F_1 = \frac{D_1}{B_1}$$

while the others are calculated recursively as:

$$E_k = -\frac{C_k}{A_k \cdot E_{k-1} + B_k} \qquad \text{and} \qquad F_k = \frac{D_k - A_k \cdot F_{k-1}}{A_k \cdot E_{k-1} + B_k}$$

At the last point ($k = kk$) the value E_{kk} is equal to zero while F_{kk} is calculated according to the above-given formula. After this, the substitution sweep is performed by the use of equation (A.2.1.4) in the descending order of the index k.

APPENDIX 2.2

Bending algorithm for reeds in steady flow

The aim of this algorithm is to calculate the deflection of a representative reed and consequently the distribution of its reduced height (effective height) over the discretisation points. The portion of the effective reed height belonging to a grid point (hr_k^n) is then used in further calculations.

The deflection of the reed is calculated according to cantilever beam theory (e.g. Timoshenko, 1955, pp.137-165). The load on the reed is a drag force caused by the flow velocity. The intensity of this load is a function of the vertical position and it is by no means constant. The drag is given by an empirical formula where it is proportional to the square of the velocity. As the flow velocity varies over the depth, the load does this also:

$$q = \frac{\rho\, CD \cdot d \cdot u^2}{2}$$

In our discrete model, the load is presented by portions of constant load derived from the representative velocity, as shown in Fig.A.2.2.1.

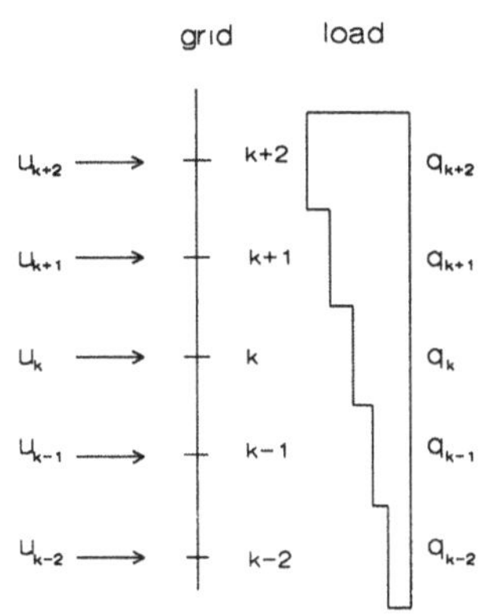

Fig.A.2.2.1

The portion of the grid between point *k-1/2* and point *k+1/2* is called the kth element. It has its load q_k, and its net deflection is $_k\delta^{net}$, while the length of the reed within it is l and consequently the height of the reed within it is hr_k. The load on the kth element is defined as:

$$q_k = \frac{\rho CD \cdot}{2} d_k \cdot (u_k)^2$$

It should at this place be pointed out that in the first instance the top of the reed coincides with a grid point but that after the first iteration for the effective reed these do not coincide any more. Thus, there is usually an element near the tip of the reed where the load belonging to that element is not spread over the whole height of the element but only along its part which corresponds to the height of the reed within that element (hr_k).

The deflection at any point along the reed is calculated by superimposing the deflections caused by each of the partial loads. For that purpose, the expression for the deflection caused by a constant load on a limited part of the beam was derived.

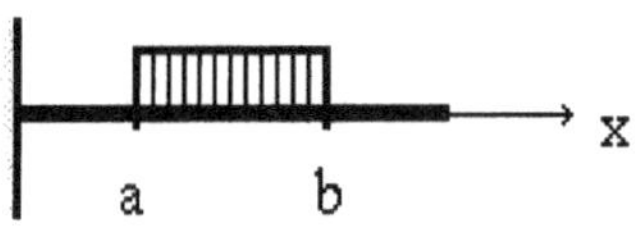

Fig.A.2.2.2

Three essentially different parts can now be distinguished on such a beam:

- a part between the fixed end and point a
- a part between points a and b
- a part between point b and the free end

The deflection δ_x at the point x caused by the load q between points a and b is given by different formulae according to the position of point x. In our model we do not need deflections for the region between points a and b; therefore, only the formulae for the other two parts are given.

for $x \leq a$

$$\delta_x^{/} = \int_a^b \frac{qx^2}{6EI}(3c - x)\,dc = \frac{qx^2}{12EI}\left(3b^2 - 3a^2 - 2x(b-a)\right)$$

for $x \geq b$

$$\delta_x^{//} = \int_a^b \frac{qc^2}{6EI}(3x - c)\,dc = \frac{q}{24EI}\left(4x(b^3 - a^3) - (b^4 - a^4)\right)$$

The deflection is calculated for points half way between two discretisation points, at the boundaries between two elements. Let us call these points i. The total deflection at point i is then expressed as:

$$\delta_i^T = \sum_{k=1}^{i} \delta'_{i,k} + \sum_{k=i+1}^{kl} \delta''_{i,k}$$

Where $\delta_{i,k}$ represents the deflection at point i caused by the load from the kth element.

The net deflection of one element is

$$\delta_k^N = \delta^T_{k+\frac{1}{2}} - \delta^T_{k-\frac{1}{2}}$$

and the length of the reed within the kth element is

$$l_k = \sqrt{\left(\delta_k^N\right)^2 + \left(\Delta z_1\right)^2}$$

The overall length (or rather height) of the reed is then distributed over the partial lengths belonging to each element. This distribution is done by advancing from the bottom with the total length being reduced for the length of bent reed belonging to each element, ($L = L - l_k$). When the remaining part of the length is shorter than the length belonging to that element, the procedure is stopped and the length of the reed belonging to this element is assigned as equal to the remaining part of the length, ($l_k = L$).

Once the effective reed height and its distribution over the discretisation grid is calculated, it should be corrected, because the effective height influences the load and the load influences the effective height. Thus, the new load is estimated, and consequently the new effective height. This procedure is schematised in Fig.A.2.2.3. This iterative procedure is repeated until a prescribed accuracy is achieved. The accuracy criterion is defined as the difference between the load height, taken as the effective height obtained from the previous calculation, and the calculated new effective height.

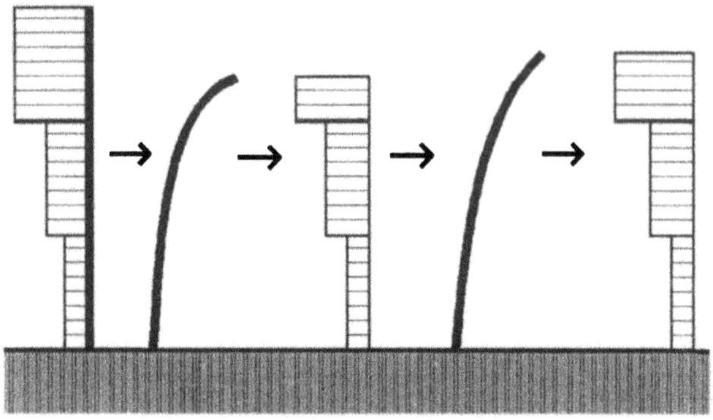

Fig.A.2.2.3

APPENDIX 3.1

Conservation equations for 1-D nearly horizontal flow

In Chapter 3 several different forms of conservation equations are used. In this appendix they will all be exposed in their full extend in order to avoid any confusion. Speaking of equations we in fact always consider a pair of equations of which one is almost invariably a mass conservation equation. The other equation in the pair is either originating from the Bernoulli equation or from the momentum equation. Moreover, each pair of equations is presented in its Eulerian form as well as in that algorithmic form which leads directly to approximation of these equations on staggered grids.

The Eulerian form of any of these pairs has a form:

$$\frac{\partial f}{\partial t} + A\frac{\partial f}{\partial x} + b = 0 \qquad \text{(A.3.1.1)}$$

while their algorithmic form is:

$$A^{-1}\frac{\partial f}{\partial t} + \frac{\partial f}{\partial x} + A^{-1}b = 0\,, \qquad |A| \neq 0 \qquad \text{(A.3.1.2)}$$

1. Equation of conservation of mass and Bernoulli equation

1.1 *Full equations*

1.1.1 Eulerian form

$$f = \begin{bmatrix} h \\ u \end{bmatrix}, \qquad A = \begin{bmatrix} u & h \\ g & u \end{bmatrix}, \qquad b = \begin{bmatrix} 0 \\ -gi + g\dfrac{u|u|}{C^2 h} \end{bmatrix}$$

$$\frac{\partial h}{\partial t} + u\frac{\partial h}{\partial x} + h\frac{\partial u}{\partial x} = 0$$

$$\frac{\partial u}{\partial t} + u\frac{\partial u}{\partial x} + g\frac{\partial h}{\partial x} - gi + g\frac{u\,|u|}{C^2 h} = 0$$

1.1.2 Algorithmic form

$$A^{-1} = \frac{1}{gh - u^2}\begin{bmatrix} -u & h \\ g & -u \end{bmatrix} \qquad A^{-1}b = \frac{1}{gh - u^2}\begin{bmatrix} -ghi + g\dfrac{u|u|}{C^2} \\ -gui - g\dfrac{u^2|u|}{C^2 h} \end{bmatrix}$$

$$-u\frac{\partial h}{\partial t} + h\frac{\partial u}{\partial t} + \left(gh - u^2\right)\frac{\partial h}{\partial x} - ghi + g\frac{u\,|u|}{C^2} = 0$$

$$g\frac{\partial h}{\partial t} - u\frac{\partial u}{\partial t} + \left(gh - u^2\right)\frac{\partial u}{\partial x} + gui - g\frac{u^2\,|u|}{h\,C^2} = 0$$

1.2 *Equations without convective term*

1.2.1 Eulerian form

$$f = \begin{bmatrix} h \\ u \end{bmatrix}, \qquad A = \begin{bmatrix} u & h \\ g & 0 \end{bmatrix}, \qquad b = \begin{bmatrix} 0 \\ -gi + g\dfrac{u|u|}{C^2 h} \end{bmatrix}$$

$$\frac{\partial h}{\partial t} + u\frac{\partial h}{\partial x} + h\frac{\partial u}{\partial x} = 0$$

$$\frac{\partial u}{\partial t} + g\frac{\partial h}{\partial x} - gi + g\frac{u|u|}{C^2 h} = 0$$

1.2.2 Algorithmic form

$$A^{-1} = \frac{1}{gh}\begin{bmatrix} 0 & h \\ g & -u \end{bmatrix} , \qquad A^{-1}b = \frac{1}{gh}\begin{bmatrix} -ghi + g\dfrac{u|u|}{C^2} \\ -gui - g\dfrac{u^2|u|}{C^2 h} \end{bmatrix}$$

$$h\frac{\partial u}{\partial t} + gh\frac{\partial h}{\partial x} - ghi + g\frac{u|u|}{C^2} = 0$$

$$g\frac{\partial h}{\partial t} - u\frac{\partial u}{\partial t} + gh\frac{\partial u}{\partial x} + gui - g\frac{u^2|u|}{hC^2} = 0$$

2. Equations of conservation of mass and momentum

2.1 *Full equations*

2.1.1 Eulerian form

$$f = \begin{bmatrix} h \\ uh \end{bmatrix} , \qquad A = \begin{bmatrix} 0 & 1 \\ gh - u^2 & 2u \end{bmatrix} , \qquad b = \begin{bmatrix} 0 \\ -ghi + g\dfrac{uh|uh|}{C^2 h} \end{bmatrix}$$

$$\frac{\partial h}{\partial t} + \frac{\partial uh}{\partial x} = 0$$

$$\frac{\partial uh}{\partial t} + \frac{\partial u^2 h}{\partial x} + gh\frac{\partial h}{\partial x} - ghi + g\frac{uh\,|uh|}{C^2 h} = 0$$

2.1.2 Algorithmic form

$$A^{-1} = \frac{1}{gh - u^2}\begin{bmatrix} -2u & 1 \\ gh - u^2 & 0 \end{bmatrix}, \qquad A^{-1}b = \frac{1}{gh - u^2}\begin{bmatrix} ghi - g\dfrac{uh\,|uh|}{C^2 h} \\ 0 \end{bmatrix}$$

$$-2u\frac{\partial h}{\partial t} + \frac{\partial uh}{\partial t} + \left(gh - u^2\right)\frac{\partial h}{\partial x} - ghi + g\frac{uh\,|uh|}{C^2 h} = 0$$

$$\frac{\partial h}{\partial t} + \frac{\partial uh}{\partial x} = 0$$

2.2 *Complete reduction of the convective momentum term*

2.2.1 Eulerian form

$$f = \begin{bmatrix} h \\ uh \end{bmatrix}, \qquad A = \begin{bmatrix} 0 & 1 \\ gh & 0 \end{bmatrix}, \qquad b = \begin{bmatrix} 0 \\ -ghi + g\dfrac{uh\,|uh|}{C^2 h} \end{bmatrix}$$

$$\frac{\partial h}{\partial t} + \frac{\partial uh}{\partial x} = 0$$

$$\frac{\partial uh}{\partial t} + gh\frac{\partial h}{\partial x} - ghi + g\frac{uh\,|uh|}{C^2 h} = 0$$

2.2.2 Algorithmic form

$$A^{-1} = \frac{1}{gh}\begin{bmatrix} 0 & 1 \\ gh & 0 \end{bmatrix} \quad , \qquad A^{-1}b = \frac{1}{gh}\begin{bmatrix} ghi - g\dfrac{uh\,|uh|}{C^2 h} \\ 0 \end{bmatrix}$$

$$\frac{\partial uh}{\partial t} + gh\frac{\partial h}{\partial x} - ghi + g\frac{uh\,|uh|}{C^2 h} = 0$$

$$\frac{\partial h}{\partial t} + \frac{\partial uh}{\partial x} = 0$$

2.3 *Partial reduction of the convective momentum term*

2.3.1 Eulerian form

$$f = \begin{bmatrix} h \\ uh \end{bmatrix} , \qquad A = \begin{bmatrix} 0 & 1 \\ gh & u \end{bmatrix} , \qquad b = \begin{bmatrix} 0 \\ -ghi + gh\dfrac{uh\,|uh|}{C^2 h^2} \end{bmatrix}$$

$$\frac{\partial h}{\partial t} + \frac{\partial uh}{\partial x} = 0$$

$$\frac{\partial uh}{\partial t} + u\frac{\partial uh}{\partial x} + gh\frac{\partial h}{\partial x} - ghi + gh\frac{uh\,|uh|}{C^2 h^2} = 0$$

2.3.2 Algorithmic form

$$A^{-1} = \frac{1}{gh}\begin{bmatrix} -u & 1 \\ gh & 0 \end{bmatrix} \quad , \qquad A^{-1}b = \frac{1}{gh}\begin{bmatrix} ghi - g\dfrac{uh\,|uh|}{C^2 h} \\ 0 \end{bmatrix}$$

$$-u\frac{\partial h}{\partial t} + \frac{\partial uh}{\partial t} + gh\frac{\partial h}{\partial x} - ghi + g\frac{uh\,|uh|}{C^2 h} = 0$$

$$\frac{\partial h}{\partial t} + \frac{\partial uh}{\partial x} = 0$$

APPENDIX 3.2

Finite difference approximations by the Abbott-Ionescu scheme

The Abbott-Ionescu scheme utilises a staggered grid as shown in Fig.A.3.2.1. The finite difference approximations of the different terms are made according to the general rules which will now be presented.

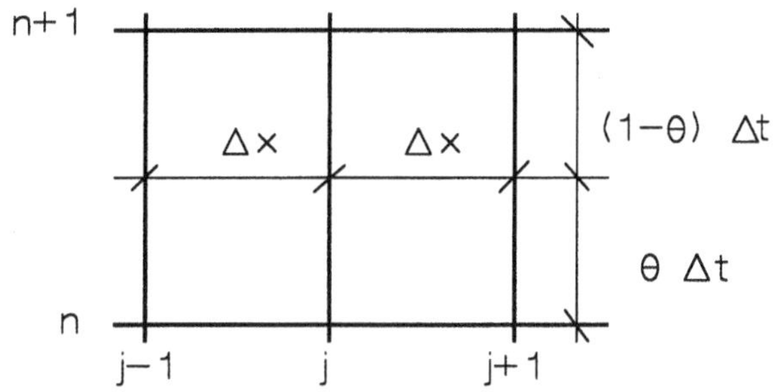

Fig.A.3.2.1

The partial derivative with respect to x:

$$\frac{\partial a}{\partial x} \approx \theta \frac{a_{j+1}^{n+1} - a_{j-1}^{n+1}}{2\Delta x} + (1-\theta)\frac{a_{j+1}^{n} - a_{j-1}^{n}}{2\Delta x}$$

The partial derivative with respect to t:

$$\frac{\partial a}{\partial t} \approx \frac{a_j^{n+1} - a_j^{n}}{\Delta t}$$

or $$\frac{\partial a}{\partial t} \approx \frac{a_{j+1}^{n+1} - a_{j+1}^{n}}{2\Delta t} + \frac{a_{j-1}^{n+1} - a_{j-1}^{n}}{2\Delta t}$$

which is dependant on the position of the variable a on the staggered grid.

The term of the type $\beta a|a|$ (primarily the friction term):

$$\beta |a| a \approx \beta_* |a_j^n| a_j^{n+1}$$

with β_* being averaged over space and time.

The term of the type

$$\frac{\partial}{\partial x}\left(\frac{a^2}{\beta}\right)$$

(like the convective momentum term):

$$\frac{\partial}{\partial x}\left(\frac{a^2}{\beta}\right) \approx \frac{1}{\beta_*}\left(\frac{a_{j+1}^{n} a_{j+1}^{n+1} - a_{j-1}^{n} a_{j-1}^{n+1}}{2\Delta x}\right)$$

The introduction of these approximations in equations of conservation in their algorithmic form, as presented in Appendix 3.1, results in a system of linear algebraic equations of the form:

$$A_j Q_{j-1}^{n+1} + B_j h_j^{n+1} + C_j Q_{j+1}^{n+1} = D_j$$

and

$$A_j h_{j-1}^{n+1} + B_j Q_j^{n+1} + C_j h_{j+1}^{n+1} = D_j$$

These two types of equations alternate throughout the system: the first one corresponds to the equation of conservation of mass and the second one either to the Bernoulli equation or to the equation of conservation of momentum. For different combination of equations, the coefficients A_j, B_j, C_j and D_j clearly lead to different expressions.

1. Equation of conservation of mass and the Bernoulli equation

1.1 *Equation of conservation of mass:*

$$A_j = \frac{\Delta t}{2\Delta x}\theta\left(u_*^2 - gh_*\right) - \frac{u_*}{2} - \frac{\Delta t g u_*}{2h_* C^2}|u_{j-1}^n|$$

$$B_j = g$$

$$C_j = -\frac{\Delta t}{2\Delta x}\theta\left(u_*^2 - gh_*\right) - \frac{u_*}{2} - \frac{\Delta t g u_*}{2h_* C^2}|u_{j+1}^n|$$

$$D_j = -\Delta t g u_* i + g h_j^n - \frac{u_*}{2}\left(u_{j+1}^n + u_{j-1}^n\right) + \frac{\Delta t}{2\Delta x}\left(u_*^2 - gh_*\right)(1-\theta)\left(u_{j+1}^n - u_{j-1}^n\right)$$

1.2 *Bernoulli equation*

1.2.1 Full equation

$$A_j = \frac{\Delta t}{2\Delta x}\theta\left(u_*^2 - gh_*\right) - \frac{u_*}{2}$$

$$B_j = h_* + g\Delta t\frac{|u_j^n|}{C^2}$$

$$C_j = -\frac{\Delta t}{2\Delta x}\theta\left(u_*^2 - gh_*\right) - \frac{u_*}{2}$$

$$D_j = \Delta t g h_* i + h_* u_j^n - \frac{u_*}{2}\left(h_{j+1}^n + h_{j-1}^n\right) + \frac{\Delta t}{2\Delta x}\left(u_*^2 - gh_*\right)(1-\theta)\left(h_{j+1}^n - h_{j-1}^n\right)$$

1.2.2 Equation without convective term

$$A_j = -\frac{\Delta t}{2\Delta x}\theta g h_*$$

$$B_j = h_* + g\Delta t\frac{|u_j^n|}{C^2}$$

$$C_j = \frac{\Delta t}{2\Delta x}\theta g h_*$$

$$D_j = \Delta t g h_* i + h_* u_j^n - \frac{\Delta t}{2\Delta x} g h_* (1-\theta)\left(h_{j+1}^n - h_{j-1}^n\right)$$

2. Equations of conservation of mass and momentum

2.1 *Equation of conservation of mass:*

$$A_j = -\frac{\theta\,\Delta t}{2\Delta x}$$

$$B_j = 1$$

$$C_j = \frac{\theta\,\Delta t}{2\Delta x}$$

$$D_j = h_j^n - \frac{(1-\theta)\,\Delta t}{2\,\Delta x}\left((uh)_{j+1}^n - (uh)_{j-1}^n\right)$$

2.2 *Equation of conservation of momentum*

2.2.1 Full equation

$$A_j = -\frac{\theta\,\Delta t}{2\,\Delta x}\left(g h_* - u_*^2\right) - u_*$$

$$B_j = 1 + \frac{g\,\Delta t}{C^2 h_*}\left|(uh)_j^n\right|$$

$$C_j = \frac{\theta\,\Delta t}{2\,\Delta x}\left(g h_* - u_*^2\right) - u_*$$

$$D_j = (uh)_j^n - \left(g h_* - u_*^2\right)\frac{(1-\theta)\,\Delta t}{2\,\Delta x}\left(h_{j+1}^n - h_{j-1}^n\right) + g\,h_*\,i\,\Delta t + u_*\left(h_{j+1}^n + h_{j-1}^n\right)$$

2.2.2 Equation with partially reduced convective term

$$A_j = -\frac{\theta\,\Delta t}{\Delta x}\,g h_* - u_*$$

$$B_j = 1 + \frac{2g\,\Delta t}{C^2 h_*}\left|(uh)_j^n\right|$$

$$C_j = \frac{\theta\,\Delta t}{\Delta x}\,g h_* - u_*$$

$$D_j = (uh)_j^n - g h_*\frac{(1-\theta)\,\Delta t}{\Delta x}\left(h_{j+1}^n - h_{j-1}^n\right) + 2g\,h_*\,i\,\Delta t + u_*\left(h_{j+1}^n + h_{j-1}^n\right)$$

2.2.3 Equation without convective term

$$A_j = -\frac{\theta\,\Delta t}{2\,\Delta x}\,g h_*$$

$$B_j = 1 + \frac{g\,\Delta t}{C^2 h_*}\left|(uh)_j^n\right|$$

$$C_j = \frac{\theta\,\Delta t}{2\,\Delta x}\,g h_*$$

$$D_j = (uh)_j^n - g h_*\,\frac{(1-\theta)\,\Delta t}{2\,\Delta x}\left(h_{j+1}^n - h_{j-1}^n\right) + g\,h_*\,i\,\Delta t$$

APPENDIX 3.3

Finite difference approximations by the Preissmann scheme

The Preissmann scheme utilises a non-staggered grid as shown in Fig.A.3.3.1. The finite difference approximations of the different terms are made according to the general rules which follow.

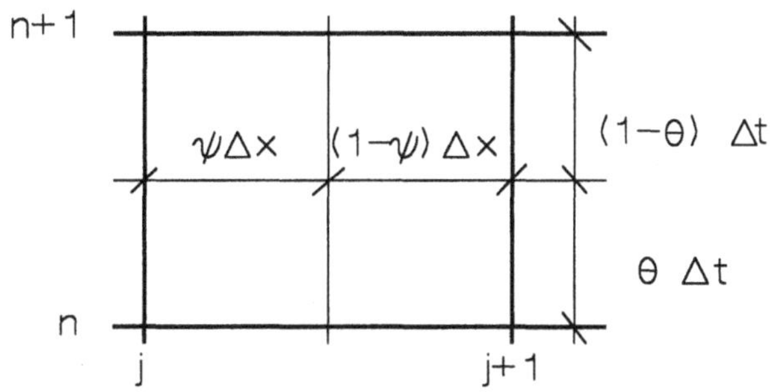

Fig.A.3.3.1

The partial derivative with respect to x:

$$\frac{\partial a}{\partial x} \approx \theta \frac{a_{j+1}^{n+1} - a_j^{n+1}}{\Delta x} + (1-\theta)\frac{a_{j+1}^{n} - a_j^{n}}{\Delta x}$$

The partial derivative with respect to t:

$$\frac{\partial a}{\partial t} \approx \psi \frac{a_{j+1}^{n+1} - a_{j+1}^{n}}{\Delta t} + (1-\psi)\frac{a_j^{n+1} - a_j^{n}}{\Delta t}$$

The term of the type $\beta\, a \,|\, a \,|$ (primarily the friction term):

$$\beta \,|a|\, a \approx \frac{\beta_*}{2}\left(|a_j^n|\, a_j^{n+1} + |a_{j+1}^n|\, a_{j+1}^{n+1} \right)$$

with β_* being the parameter β averaged in space and time.

The term of the type

$$\frac{\partial}{\partial x}\left(\frac{a^2}{\beta}\right)$$

(like the convective momentum term):

$$\frac{\partial}{\partial x}\left(\frac{a^2}{\beta}\right) \approx \frac{1}{\beta_*}\left(\frac{a_{j+1}^n a_{j+1}^{n+1} - a_j^n a_j^{n+1}}{\Delta x}\right)$$

The introduction of these approximations in the equations of conservation in their algorithmic form, as presented in Appendix 3.1, results in a system of linear algebraic equations of the form:

$$A_j Q_j^{n+1} + B_j h_j^{n+1} + C_j Q_{j+1}^{n+1} + D_j h_{j+1}^{n+1} = E_j$$

For the different equations, the coefficients A_j, B_j, C_j , D_j and E_j are again clearly different.

1. Equation of conservation of mass and the Bernoulli equation

1.1 *Equation of conservation of mass:*

$$A_j = -h_* \theta$$

$$B_j = \frac{\Delta x}{\Delta t}(1-\psi) - u_*\theta$$

$$C_j = h_*\theta$$

$$D_j = \frac{\Delta x}{\Delta t}\psi + u_*\theta$$

$$E_j = \frac{\Delta x}{\Delta t}\left(\psi\, h_{j+1}^n + (1-\psi)\, h_j^n\right) - h_*(1-\theta)\left(u_{j+1}^n - u_j^n\right) - u_*(1-\theta)\left(h_{j+1}^n - h_j^n\right)$$

1.2 *Bernoulli equation*

1.2.1 Full equation

$$A_j = \frac{\Delta x}{\Delta t}(1-\psi) - u_*\theta + \frac{g\,\Delta x}{2\,h_*\,C^2}\left|u_j^n\right|$$

$$B_j = -g\,\theta$$

$$C_j = \frac{\Delta x}{\Delta t}(1-\psi) + u_*\theta + \frac{g\,\Delta x}{2\,h_*\,C^2}\left|u_j^n\right|$$

$$D_j = g\,\theta$$

$$E_j = \frac{\Delta x}{\Delta t}\left(\psi\, u_{j+1}^n + (1-\psi)\, u_j^n\right) - u_*(1-\theta)\left(u_{j+1}^n - u_j^n\right) - g\,(1-\theta)\left(h_{j+1}^n - h_j^n\right) + g\,i\,\Delta x$$

1.2.2 Equation without convective term

$$A_j = \frac{\Delta x}{\Delta t}(1-\psi) + \frac{g\,\Delta x}{2\,h_*\,C^2}\left|u_j^n\right|$$

$$B_j = -g\,\theta$$

$$C_j = \frac{\Delta x}{\Delta t}(1-\psi) + \frac{g\,\Delta x}{2\,h_*\,C^2}\left|u_j^n\right|$$

$$D_j = g\,\theta$$

$$E_j = \frac{\Delta x}{\Delta t}\left(\psi\,u_{j+1}^n + (1-\psi)\,u_j^n\right) - u_*(1-\theta)\left(u_{j+1}^n - u_j^n\right) - g(1-\theta)\left(h_{j+1}^n - h_j^n\right) + g\,i\,\Delta x$$

2. Equations of conservation of mass and momentum

2.1 *Equation of conservation of mass:*

$$A_j = -\theta$$

$$B_j = \frac{\Delta x}{\Delta t}(1-\psi)$$

$$C_j = \theta$$

$$D_j = \frac{\Delta x}{\Delta t}\psi$$

$$E_j = \frac{\Delta x}{\Delta t}\left(\psi\, h_{j+1}^n + (1-\psi)\, h_j^n\right) - (1-\theta)\left((uh)_{j+1}^n - (uh)_j^n\right)$$

2.2 *Equation of conservation of momentum*

2.2.1 Full equation

$$A_j = \frac{\Delta x}{\Delta t}(1-\psi) - \frac{(uh)_j^n}{h_*} + \frac{g\,\Delta x}{2\,h_*\,C^2}\left|(uh)_j^n\right|$$

$$B_j = -g\,h_*\,\theta$$

$$C_j = \frac{\Delta x}{\Delta t}\psi - \frac{(uh)_{j+1}^n}{h_*} + \frac{g\,\Delta x}{2\,h_*\,C^2}\left|(uh)_{j+1}^n\right|$$

$$D_j = g\,h_*\,\theta$$

$$E_j = \frac{\Delta x}{\Delta t}\left(\psi\,(uh)_{j+1}^n + (1-\psi)\,(uh)_j^n\right) - g\,h_*\,(1-\theta)\left(h_{j+1}^n - h_j^n\right) + g\,h_*\,i\,\Delta x$$

2.2.2 Equation with partially reduced convective term

$$A_j = \frac{\Delta x}{\Delta t}(1-\psi) - u_* \theta + \frac{g \Delta x}{2 h_* C^2}\left|(uh)_j^n\right|$$

$$B_j = -g h_* \theta$$

$$C_j = \frac{\Delta x}{\Delta t}\psi - \frac{(uh)_{j+1}^n}{h_*} + \frac{g \Delta x}{2 h_* C^2}\left|(uh)_{j+1}^n\right|$$

$$D_j = g h_* \theta$$

$$E_j = \frac{\Delta x}{\Delta t}\left(\psi (uh)_{j+1}^n + (1-\psi)(uh)_j^n\right) - g h_* (1-\theta)\left(h_{j+1}^n - h_j^n\right) - u_* (1-\theta)\left((uh)_{j+1}^n - (uh)_j^n\right) + g h_* i$$

2.2.3 Equation without convective term

$$A_j = \frac{\Delta x}{\Delta t}(1-\psi) + \frac{g \Delta x}{2 h_* C^2}\left|(uh)_j^n\right|$$

$$B_j = -g h_* \theta$$

$$C_j = \frac{\Delta x}{\Delta t}\psi + \frac{g \Delta x}{2 h_* C^2}\left|(uh)_{j+1}^n\right|$$

$$D_j = g h_* \theta$$

$$E_j = \frac{\Delta x}{\Delta t}\left(\psi (uh)_{j+1}^n + (1-\psi)(uh)_j^n\right) - g h_* (1-\theta)\left(h_{j+1}^n - h_j^n\right) + g h_* i \Delta x$$

APPENDIX 3.4

The double sweep algorithm

The double sweep algorithm is the most economical purely-sequential method for the solution of systems of linear equations characterised by banded matrices. However, depending on the band-width, the algorithm differs slightly. The finite difference approximation by the Abbot-Ionescu scheme results in a tri-diagonal matrix while the Preissmann scheme produces a penta-diagonal matrix, even though this can, of course, always be reduced further to a tri-diagonal form. For reasons of completeness and associated generality, both versions of the double-sweep algorithm are exposed.

1. Tri-diagonal matrices

The recurrence relation in the form

$$a_j^{n+1} = E_j \cdot a_{j+1}^{n+1} + F_j$$

is defined where a_j^{n+1} stands for any unknown (u or h)

In the elimination sweep, the coefficients E_j and F_j are calculated recursively as:

$$E_j = \frac{-C_j}{A_j E_{j-1} + B_j} \qquad \text{and} \qquad F_j = \frac{D_j - A_j F_{j-1}}{A_j E_{j-1} + B_j}$$

The elimination sweep is initiated by setting the values of the initial coefficients E_j and F_j according to one boundary condition. When the last recurrence relation is calculated, the other boundary condition is employed in order to get the value of the unknown at that point. After this, the substitution sweep, in which previously calculated recurrence relations are used, is performed.

2. Penta-diagonal matrices

The procedure of the double sweep is equivalent to that just explained above for tri-diagonal matrices; only the recurrence relations and the expressions for the calculation of these coefficients differ.

The recurrence relations now have the form

$$h_j^{n+1} = H_j \cdot u_{j+1}^{n+1} + I_j \cdot h_{j+1}^{n+1} + J_j$$

$$u_j^{n+1} = F_j \cdot h_{j+1}^{n+1} + G_j$$

In the elimination sweep the coefficients H_j, I_j, J_j, F_j and G_j are calculated recursively as:

$$H_j = \frac{-C1_j}{A1_j F_{j-1} + B1_j}, \qquad I_j = \frac{-D1_j}{A1_j F_{j-1} + B1_j},$$

$$J_j = \frac{E1_j - A1_j G_{j-1}}{A1_j F_{j-1} + B1_j}, \qquad F_j = \frac{\left(A2_j F_{j-1} + B2_j\right) I_j + D2_j}{\left(A2_j F_{j-1} + B2_j\right) H_j + C2_j},$$

$$G_j = \frac{-\left(A2_j F_{j-1} + B2_j\right) J_j + E2_j - A2_j G_{j-1}}{\left(A2_j F_{j-1} + B2_j\right) H_j + C2_j}$$

where the coefficients *A1, B1, C1, D1* and *E1* denote coefficients from the equations of conservation of mass and the coefficients *A2, B2, C2, D2* and *E2* those arising from either Bernoulli equations or from equations of conservation of momentum.

APPENDIX 3.5

Stability and accuracy of the finite difference schemes

A linearised analysis by Fourier series expansions of the Preissmann scheme applied on the momentum equation with partially and fully-reduced convective momentum term is first presented.

The de Saint Venant equation for a rectangular, horizontal, frictionless channel with reduced convective momentum term has the form:

$$\frac{\partial Q}{\partial x} + b\frac{\partial h}{\partial t} = 0$$

$$\frac{\partial Q}{\partial t} + \nu u\frac{\partial Q}{\partial x} + gA\frac{\partial h}{\partial x} = 0 \qquad (A.3.5.1)$$

where ν is a reduction coefficient which, by being zero, provides the full reduction of the convective momentum term, and which, by taking a value of unity, introduces the partially reduced convective momentum term described in the main text.

After approximation of the partial derivatives by finite differences according to the Preissmann scheme and replacement of each variable by its Fourier series expansion, only the kth component of the expansion is considered (See Abbott 1979, p.168). The amplification coefficient ϕ is defined for any variable as:

$$\phi = \frac{\xi_k^{n+1}(a)}{\xi_k^{n}(a)} \quad ;$$

where ξ_k^n is the Fourier amplitude coefficient of the kth dimensionless wave number at time Δt. The following equations are then obtained:

$$b\frac{\phi-1}{\Delta t}\left(\cos\alpha'+(2\psi-1)i\sin\alpha'\right)\xi_k^n(h) + \frac{\theta(\phi-1)+1}{\Delta x}2i\sin\alpha'\,\xi_k^n(Q) = 0$$

$$gA\frac{\theta(\phi-1)+1}{\Delta x}2i\sin\alpha'\,\xi_k^n(h) +$$

$$+\left[\frac{\phi-1}{\Delta t}\left(\cos\alpha'+(2\psi-1)i\sin\alpha'\right) + \nu u\frac{\theta(\phi-1)+1}{\Delta x}2i\sin\alpha'\right]\xi_k^n(Q) = 0$$

where $\alpha' = \pi/N_x$ with N_x defined as the number of points per wave length.

Such a system of linear homogenous equations has a non-trivial set of solutions if and only if the matrix of its coefficients is singular. Following this condition, but simplifying for convenience to the fully centred scheme with θ=0.5 and ψ=0.5, we get:

$$b\frac{\phi-1}{\Delta t}\cos\alpha'\left(\frac{\phi-1}{\Delta t}\cos\alpha' + \nu\frac{\phi+1}{\Delta x}iu\sin\alpha'\right) + \frac{(\phi+1)^2}{\Delta x^2}gA\sin^2\alpha' = 0$$

After rearranging, this becomes:

$$(\phi-1)^2 + \nu iu\frac{\Delta t}{\Delta x}\tan\alpha'(\phi-1)(\phi+1) + gh\frac{\Delta t^2}{\Delta x^2}\tan^2\alpha'(\phi+1)^2 = 0 \qquad \text{(A.3.5.2)}$$

The Courant number (Cr) and the Froude number (Fr) are substituted in the above expression together with a new variable χ as defined by:

$$\chi = \frac{\phi-1}{\phi+1}, \qquad Fr = \frac{u}{\sqrt{gh}}, \qquad Cr = \left(u+\sqrt{gh}\right)\frac{\Delta t}{\Delta x}$$

Equation (A.3.5.2) then takes the form:

$$\chi^2 + \nu\, i \tan\alpha' \frac{Cr\,Fr}{Fr+1}\,\chi + \tan^2\alpha' \frac{Cr^2}{(Fr+1)^2} = 0$$

We then find, for the fully-reduced convective momentum term with $\nu = 0$:

$$\chi_{1,2} = \pm i\,a$$

and, for the partial reduction with $\nu = 1$:

$$\chi_{1,2} = -i\,a \left(\frac{Fr}{2} \pm \sqrt{\frac{Fr^2}{4} + 1} \right) , \qquad \text{with} \qquad a = \tan\alpha' \frac{Cr}{Fr+1}$$

As, in both cases, the variable χ takes a purely imaginary value, it follows from the definition of ϕ that $|\phi| = 1$, i.e. there is no amplification of any wave-number component between two consecutive time levels for the cases considered. The scheme thus appears to be unconditionally stable for both cases.

The phase error is presented by the 'relative celerity' (RC) defined as the ratio between the numerical and the continuum celerity. The numerical celerity is equal to the argument of the amplification factor while the continuum celerity is defined by the slope of the C^+ characteristic of the full de Saint Venant equation.

Hence, the expressions for the relative celerity in the two cases considered are (Abbott and Basco, 1989, p.86):

$$RC = \frac{-\arctan\dfrac{-2a}{1-a^2}}{Cr\,\dfrac{2\pi}{Nx}} \qquad \text{for the fully-reduced convective momentum term.}$$

These expressions have been used to construct the phase portraits presented in Fig.A.3.5.1.

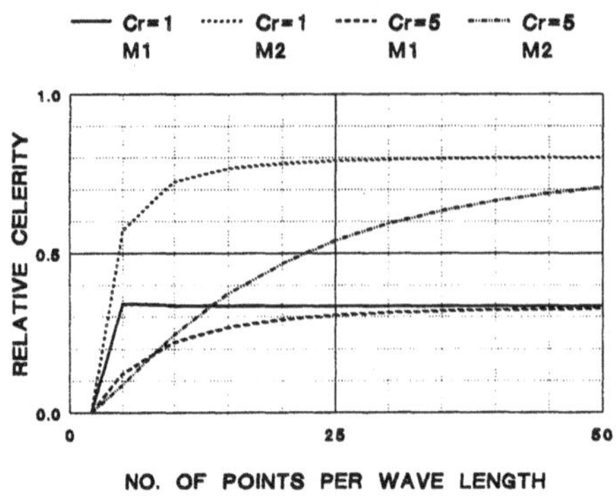

a) Fr = 2

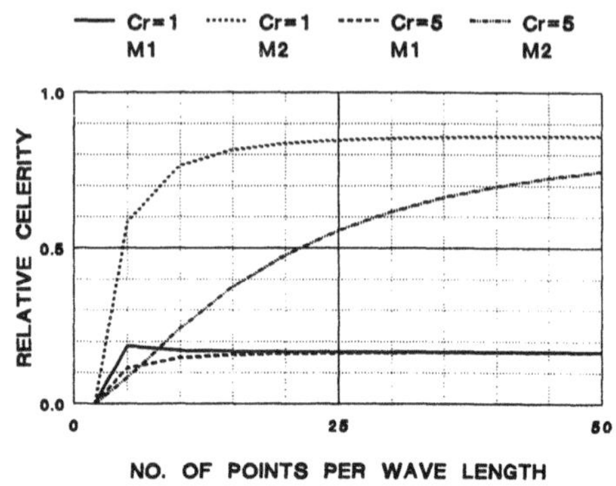

b) Fr = 5

Fig.A.3.5.1 The phase portraits for the Preissmann scheme applied on the mass equation with the fully (M1) and partially (M2) reduced convective term

$$RC = \frac{-\arctan \dfrac{-2ac}{1-a^2c^2}}{Cr\ \dfrac{2\pi}{Nx}}$$ for the partially-reduced convective momentum term

with: $$c = \frac{Fr}{2} \pm \sqrt{\frac{Fr^2}{4} + 1}$$

The phase portraits presented in Fig.a.35.1 show clearly the difference between the performance of the models based on the full (M1) or partial (M2) reduction of the convective momentum term. Whereas the relative celerity of the case with partial reduction reaches approximately 0.8 for more than 20 points per wave length (Courant number equal to one) in the other case it is far below this value. Moreover, with an increase of the Froude number, the celerity ratio of the M2 model improves slightly while, on the other hand, the celerity ratio of the M1 model deteriorates drastically.

These observations are fully in accordance with the results obtained earlier for the celerity coefficient as presented in Fig.A.3.5.1 : they demonstrate, once more, the unsuitability of the M1 model, even though it does provide unconditionally stable 'results'.

The phase portraits for the Abbott-Ionescu scheme, presented in the Fig.A.3.5.2, are obtained as a result of a similar stability analysis to that given for the Preissmann scheme. All the properties observed in the case of Fig.A.3.5.1 can, as well, be recognised here. Hence, it can be concluded that the phase errors of both schemes, when applied to the supercritical flow simulation, have similar characteristics.

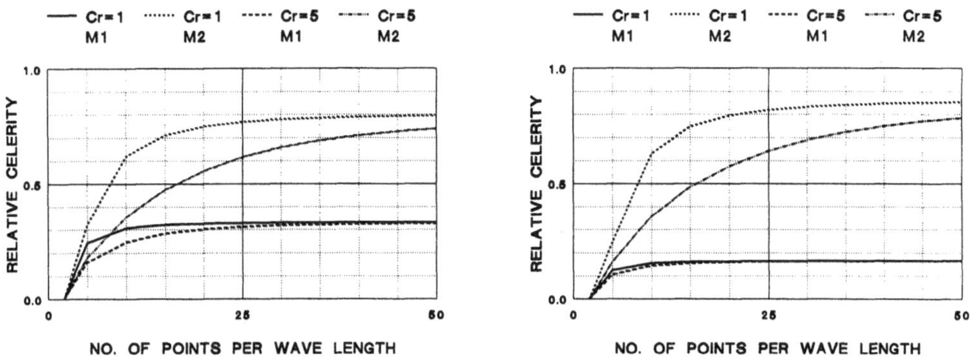

a) Fr = 2 b) Fr = 5

Fig.A.3.5.2 The phase portraits for the Abbott-Ionescu scheme applied on the mass equation with fully (M1) and partially (M2) reduced convective momentum term

APPENDIX 3.6

Stability of the algorithmic structure

This stability analysis is directed towards the algorithmic structure used in the actual computation. The idea is to analyse the performance of the double-sweep algorithm used as a solution method for the system of equations obtained after translation from the continuum equation to the finite difference schemes for both the subcritical and supercritical flow conditions. This is not, of course, an analysis of stability in the Richtmyer-von Neuman sense.

The double-sweep method is a solution algorithm for bounded matrices based on the Gauss elimination method. In a strict algebraic sense (Volkov, 1986, pp.155-160), a *sufficient* condition for the existence of a unique solution is that the principle diagonal is dominant. This to say that if, for example, we can write the jth row of a tri-diagonal matrix as

$$A_j x_{j-1} + B_j x_j + C_j x_{j+1} = D_j\,, \qquad j=1,\dots,jj \tag{A.3.6.1}$$

where x is a generalised unknown, which may conventionally be any of h or u or uh or any other such dependent variable, and

$$|B_j| \geq |A_j| + |C_j| \geq |A_j| > 0$$

then the matrix has a dominating principal diagonal and the double sweep algorithm will be well-conditioned and so stable.

Unfortunately, as observed earlier, the matrices of coefficients that arise here do not fulfil this condition: even in the case of subcritical flow, where we do not doubt the stability of the double-sweep algorithm, we do not necessarily obtain a matrix with a dominant principal diagonal. It is then clear that the above condition, although sufficient, may not always be necessary. Hence, it is not entirely relevant in this context and a weaker condition has to be found.

In the case of the Abbott-Ionescu scheme, we deal with a tri-diagonal matrix, while for the

Preissmann scheme we have, at least initially, a penta-diagonal matrix even though this can easily be reduced to a tri-diagonal form (Abbott and Basco, 1989, pp.242-244). Here, for simplicity of exposition, only the tri-diagonal case will be examined.

Within the double sweep algorithm, the solution at any space point j is obtained from the solution in a point $j+1$ following the linear expression:

$$x_j = E_j x_{j+1} + F_j$$

where E_j, and F_j can be obtained from the coefficients A_j, B_j, C_j, and D_j by substitution in (A.3.6.1).

When for x_{j+1} a similar relation is substituted and the same procedure is repeated m times, we arrive at the expression:

$$x_j^{n+1} = \left(\prod_{k=0}^{m-1} E_{j+k} \right) x_{j+m}^{n+1} + \sum_{l=0}^{m-1} \left(\frac{\prod_{k=0}^{l} E_{j+k}}{E_{j+l}} F_{j+l} \right)$$

From this expression, it can be concluded that if $\mid E_j \mid > 1$ then we can always take m as large as we wish to obtain:

$$\prod_{k=0}^{m-1} E_{j+k} > M \tag{A.3.6.2}$$

for any M, so that the solution for x_j is not uniformly bounded. It is thus unstable: the results will amplify along the space variable. Accordingly, in order to demonstrate the absence of spatial amplification within the double-sweep algorithm, the condition:

$$\left| E_j \right| \leq 1 \quad , \quad \forall j.$$

should be fulfilled. However, this condition is really nothing more than a rewriting of dominant-principal-diagonal condition, which, as already seen, need not necessarily be fulfilled.

As mass and momentum equations are always alternated in the tri-diagonal form of the Abbott-Ionescu scheme, the idea arises to couple them by observing that the product:

$$\prod_{k=0}^{m-1} E_{j+k} \quad \text{can be rewritten as} \quad \prod_{i=0,2,4,..}^{m-1} \left(E_{j+i} \times E_{j+i+1} \right)$$

From the above expression it is clear that if $| E_j E_{j+1} | > 1$, an m can be found such that condition (A.3.6.2) is fulfilled for any M (i.e. the solution is not uniformly bounded, i.e. it is unstable). Accordingly, the condition

$$|E_j E_{j-1}| \leq 1 \quad , \quad \forall j. \tag{A.3.6.3}$$

ensures that there is no spatial amplification: the solution is then uniformly bounded.

The coefficient E_j is defined by recursion, however, as:

$$E_j = \frac{-C_j}{A_j E_j + B_j}$$

After it is substituted in condition (A.3.6.3), we obtain:

$$\left| \frac{-C_j E_{j-1}}{A_j E_{j-1} + B_j} \right| \leq 1$$

or

$$| -C_j E_{j-1} | \leq | A_j E_{j-1} + B_j | \leq | A_j E_{j-1} | + | B_j |$$

This provides

$$| -C_j E_{j-1} | - | A_j E_{j-1} | \leq | B_j |$$

or

$$| C_j | - | A_j | \leq \left| \frac{B_j}{E_{j-1}} \right| \tag{A.3.6.4}$$

In the case where

$$|C_j| \leq |A_j| \Rightarrow |C_j| - |A_j| \leq 0 \quad ,$$

this would be sufficient for fulfilling the condition (A.3.6.3) because the right hand side of the condition (A.3.6.4) is, by the definition of the absolute value, greater than or equal to zero.

When the mass equation and the full Bernoulli equation are used, coefficients A_j and C_j from each of the equations have the same formal expressions:

$$A_j = -u + (u^2 - gh)\frac{\Delta t}{2\Delta x} \quad , \qquad C_j = -u - (u^2 - gh)\frac{\Delta t}{2\Delta x}$$

For **subcritical** flow conditions, $u^2 - gh < 0$, and from the above-given formulae it follows that:

$$|C_j| < |A_j| \quad \text{and} \quad |C_j| - |A_j| < 0 \quad ,$$

which is to say that the condition (A.3.6.3) is fulfilled for every j and the solution is stable.

For **supercritical** flow conditions, $u^2 - gh > 0$, so that from the expressions for A_j and C_j it is evident that

$$|C_j| > |A_j| \quad \text{and} \quad |C_j| - |A_j| > 0 \quad ,$$

Unfortunately, from these conditions, nothing can be concluded about the satisfying of the condition (A.3.6.3). On the other hand, when the Bernoulli equation without the convective momentum term is used, the expressions for A_j and C_j differ between the mass and the Bernoulli equations as follows:

mass equation:

$$A_j = -u - gh\frac{\Delta t}{2\Delta x} \quad , \quad C_j = -u + gh\frac{\Delta t}{2\Delta x}$$

giving

$$|C_j| < |A_j| \quad \text{and} \quad |C_j| - |A_j| < 0 \quad ,$$

Bernoulli equation:

$$A_j = -gh\frac{\Delta t}{2\Delta x} \quad , \quad C_j = gh\frac{\Delta t}{2\Delta x}$$

giving

$$|C_j| = |A_j| \quad \text{and} \quad |C_j| - |A_j| = 0 \quad ,$$

For both equations it is obvious that condition (A.3.6.3) is fulfilled independently of the flow conditions. Thus, it can be concluded that in the case of Bernoulli equation without the convective term the solution of the double-sweep algorithm is unconditionally stable.

In the case of the Preissmann scheme, the original matrix of the equation system is penta-diagonal and recurrence relations for its solution by the double-sweep algorithm are given by (Abbott and Basco, 1989, pp.242-244):

$$h_{j+1} = F_j u_{j+1} + G_j$$

$$u_j = H_j h_{j+1} + I_j u_{j+1} + J_j$$

When similar relations are substituted for u_{j+1} and h_{j+1} and the same procedure is repeated m times, the expression obtained is:

$$u_j = \prod_{k=0}^{m-1} \left(H_{j+k} F_{j+k} + I_{j+k} \right) u_{j+m} + \ldots..$$

From this expression it can be concluded that if $\left| F_j H_j + I_j \right| > 1$, m can be chosen as large as we wish to obtain:

$$\prod_{k=0}^{m-1} \left| F_{j+k} H_{j+k} + I_{j+k} \right| > M$$

for any M, so that the solution for u_j is not uniformly bounded, and is thus again unstable.

Accordingly, in order to ensure stability of the double-sweep algorithm for the Preissmann scheme, the condition:

$$\left| F_j H_j + I_j \right| \leq 1 \qquad , \forall j \tag{A.3.6.5}$$

should be fulfilled. We observe that in the condition (A.3.6.5), pairs of equations are included; these are equations of conservation of mass and conservation of momentum applied on the same 'box'. In the case of Abbott-Ionescu scheme, the same two equations, but applied to two neighbouring points, were used to obtain condition (A.3.6.3).

APPENDIX 4.1

Finite difference approximations of the de Saint Venant equations

Flows in the open-channel networks are described by the de Saint Venant equations for one-dimensional free-surface flow:

Continuity:
$$\frac{\partial Q}{\partial x} + b_s \frac{\partial h}{\partial t} = 0 \tag{A.4.1.1}$$

Momentum:
$$\frac{\partial Q}{\partial t} + \frac{\partial}{\partial x}\left(\beta \frac{Q^2}{A}\right) + gA\left(\frac{\partial h}{\partial x} - I_b\right) + gA\frac{Q|Q|}{K^2} = 0 \tag{A.4.1.2}$$

where:

Q	- discharge	g	- gravity acceleration
h	- water depth	β	- Boussinesq coefficient
b_s	- storage width	K	- Conveyance ($K=CA\sqrt{R}$)
A	- cross-section area	C	- Chezy resistance coefficient
I_b	- bottom slope	R	- hydraulic radius

In order to approximate these equations on staggered grid following the Abbott-Ionescu finite difference scheme, the convective term has to be changed. We have:

$$\frac{\partial}{\partial x}\left(\beta \frac{Q^2}{A}\right) = \frac{\beta\, 2Q}{A}\frac{\partial Q}{\partial x} - \frac{\beta\, Q^2}{A^2}\frac{\partial A}{\partial x}$$

From the continuity equation (A.4.1.1)

$$\frac{\partial Q}{\partial x} = - b_s \frac{\partial h}{\partial t}$$

while if we assume that the cross-section area is a function of the water depth $A = f(h)$

$$\frac{\partial A}{\partial x} = \frac{dA}{dh}\frac{\partial h}{\partial x} \approx b_T \frac{\partial h}{\partial x}$$

Thus the convective momentum term can be replaced by

$$\frac{\partial}{\partial x}\left(\beta \frac{Q^2}{A}\right) = - \frac{2\beta b_s Q}{A}\frac{\partial h}{\partial t} - \frac{\beta Q^2 b_T}{A^2}\frac{\partial h}{\partial x}$$

Finally, the equations to be approximated are:

$$\frac{\partial Q}{\partial x} + b_s \frac{\partial h}{\partial t} = 0$$

$$\frac{\partial Q}{\partial t} - \frac{2\beta b_s Q}{A}\frac{\partial h}{\partial t} + \left(gA - \frac{\beta Q^2 b_T}{A^2}\right)\frac{\partial h}{\partial x} - gA - i_b + g$$

Finite difference approximations:

Continuity equation

$$\frac{\partial Q}{\partial x} \approx \frac{\theta}{2\Delta x}\left(Q_{j+1}^{n+1} - Q_{j-1}^{n+1}\right) + \frac{(1-\theta)}{2\Delta x}\left(Q_{j+1}^{n} - Q_{j-1}^{n}\right)$$

$$b_s \frac{\partial h}{\partial t} \approx b_{s_j}^{n+\frac{1}{2}} \frac{h_j^{n+1} - h_j^n}{\Delta t}$$

Upon substitution of these two approximations into the continuity equation, it can be rewritten in the form of a linear algebraic equation as:

$$A_j Q_{j-1} + B_j h_j + C_j Q_{j+1} = D_j$$

with coefficients:

$$A_j = -\frac{\theta}{2\Delta x} \quad , \quad B_j = -\frac{b_{s_j}^{n+1/2}}{\Delta t} \quad , \quad C_j = \frac{\theta}{2\Delta x}$$

$$D_j = \frac{b_{s_j}^{n+1/2}}{\Delta t} h_j^n - \frac{(1-\theta)}{2\Delta x}\left(Q_{j+1}^n - Q_{j-1}^n\right)$$

Momentum equation

$$\frac{\partial Q}{\partial t} \approx \frac{Q_j^{n+1} - Q_j^n}{\Delta t}$$

$$\frac{2\beta b_s Q}{A}\frac{\partial h}{\partial t} \approx \frac{\beta b_{s_j}^{n+1/2} Q_j^{n+1/2}}{A_j^{n+1/2}} \left(\frac{h_{j+1}^{n+1} - h_{j+1}^n}{\Delta t} + \frac{h_{j-1}^{n+1} - h_{j-1}^n}{\Delta t} \right)$$

$$\left(gA - \frac{\beta b_T Q^2}{A^2} \right) \frac{\partial h}{\partial x} \approx \left(gA_j^{n+1/2} - \frac{\beta b_{T_j}^{n+1/2} \left(Q_j^{n+1/2}\right)^2}{\left(A_j^{n+1/2}\right)^2} \right) \left(\frac{\theta}{2\Delta x}\left(h_{j+1}^{n+1} - h_{j-1}^{n+1}\right) + \frac{(1-\theta)}{2\Delta x}\left(h_{j+1}^n - h_{j-1}^n\right) \right)$$

$$gAi \approx gA_j^{n+1/2} i$$

$$g A \frac{Q\,|Q|}{K^2} \approx g A_j^{n+1/2} \frac{Q_j^{n+1} \left|Q_j^n\right|}{\left(K_j^{n+1/2}\right)^2}$$

Upon substitution of these approximations into the momentum equation, it can be rewritten in the form of a linear algebraic equation as:

$$A_j h_{j-1} + B_j Q_j + C_j h_{j+1} = D_j$$

with coefficients:

$$A_j = - \frac{\beta Q_j^{n+1/2} b_{s_j}^{n+1/2}}{\Delta t A_j^{n+1/2}} - \left(g A_j^{n+1/2} - \frac{\beta \left(Q_j^{n+1/2}\right)^2 b_{T_j}^{n+1/2}}{\left(A_j^{n+1/2}\right)^2} \right) \frac{\theta}{2 \Delta x}$$

$$B_j = \frac{1}{\Delta t} + g A_j^{n+1/2} \frac{\left|Q_j^n\right|}{\left(K_j^{n+1/2}\right)^2}$$

$$C_j = - \frac{\beta Q_j^{n+1/2} b_{s_j}^{n+1/2}}{\Delta t A_j^{n+1/2}} + \left(g A_j^{n+1/2} - \frac{\beta \left(Q_j^{n+1/2}\right)^2 b_{T_j}^{n+1/2}}{\left(A_j^{n+1/2}\right)^2} \right) \frac{\theta}{2 \Delta x}$$

$$D_j = \frac{Q_j^n}{\Delta t} - \frac{\beta Q_j^{n+1/2} b_{s_j}^{n+1/2}}{\Delta t A_j^{n+1/2}} \left(h_{j+1}^n + h_{j-1}^n\right) -$$

$$- \left(g A_j^{n+1/2} - \frac{\beta \left(Q_j^{n+1/2}\right)^2 b_{T_j}^{n+1/2}}{\left(A_j^{n+1/2}\right)^2} \right) \frac{1-\theta}{2 \Delta x} \left(h_{j+1}^n - h_{j-1}^n\right) + g A_j^{n+1/2} i$$

For the rectangular cross-section used in the test version of the generalised solution method, these coefficients are reduced to:

$$A_j = -\frac{\beta Q_j^{n+1/2}}{\Delta t\, h_j^{n+1/2}} - \left(g A_j^{n+1/2} - \frac{\beta \left(Q_j^{n+1/2}\right)^2}{A_j^{n+1/2}\, h_j^{n+1/2}} \right) \frac{\theta}{2\,\Delta x}$$

$$B_j = \frac{1}{\Delta t} + g \frac{\left|Q_j^n\right| \left(b + 2 h_j^{n+1/2}\right)}{\left(C A_j^{n+1/2}\right)^2}$$

$$C_j = -\frac{\beta Q_j^{n+1/2}}{\Delta t\, h_j^{n+1/2}} + \left(g A_j^{n+1/2} - \frac{\beta \left(Q_j^{n+1/2}\right)^2}{A_j^{n+1/2}\, h_j^{n+1/2}} \right) \frac{\theta}{2\,\Delta x}$$

$$D_j = \frac{Q_j^n}{\Delta t} - \frac{\beta Q_j^{n+1/2}}{\Delta t\, h_j^{n+1/2}} \left(h_{j+1}^n + h_{j-1}^n\right) -$$

$$- \left(g A_j^{n+1/2} - \frac{\beta \left(Q_j^{n+1/2}\right)^2}{A_j^{n+1/2}\, h_j^{n+1/2}} \right) \frac{1-\theta}{2\,\Delta x} \left(h_{j+1}^n - h_{j-1}^n\right) + g A_j^{n+1/2} i$$

Boundary and internal conditions

At each end-vertex a boundary condition is expected. It can be either the water depth or the discharge expressed as a function of time. The detailed treatment of the boundary conditions is explained in Appendix 4.3. Moreover, at each vertex an external discharge can be specified which enters into the continuity equation at that particular vertex.

The continuity equations at all the vertices have the form:

$$\sum^{d_i} Q_e + q = 0$$

where:
- i - number of the vertex;
- d_i - degree of the ith vertex;
- e - indices of all the edges incident to this vertex;
- Qe - discharge in edges incident to the ith vertex;
- q - external discharge;

Beside these continuity equations, other compatibility conditions obtained at vertices of the network may be expressed to some approximation as equal water levels. At each vertex of the network graph there is an unknown denoted by Y_i, $i = 1$ to n, representing the water levels at the vertices. This water level is common for all incident edges. The set of unknowns along each edge,(which are discharges and water levels alternately) is denoted by X_j, $j=1$ to the number of grid points along the kth edge (more usually denoted by jj). In this manner the system of equations along each edge can be seen to have the form:

$$\begin{bmatrix} * & * & * & & & \\ & * & * & * & & \\ & & * & * & * & \\ & & & * & * & * \\ & & & & \cdot & \cdot & \cdot \end{bmatrix} \times \begin{bmatrix} Y_i \\ X_1 \\ X_2 \\ X_3 \\ \cdot \end{bmatrix} = \begin{bmatrix} * \\ * \\ * \\ * \\ \cdot \end{bmatrix}$$

In following Appendices 4.2 untill 4.4, a solution method for the set of all these equations is elaborated.

APPENDIX 4.2

Algorithms for the definition of the control sequence

The algorithms for the definition of the control sequence are part of the generalised solution method for flow conditions in channel networks. They are based on graph theoretical presentation of channel networks such as is extremely well suited for storage in computer memory.

Any network is defined by its incidence matrix as an undirected graph. Basically, three different algorithms are applied in sequence on an incidence matrix of a network graph. The final result of these algorithms is an incidence matrix of a di-graph and a set of lists defining the order of edges and vertices to be treated by the generalised solution method. The orientation of the graph edges corresponds to a numbering of the discretisation points within each edge in order to comply with the same set of equations for elimination.

These three algorithms are:

- *splitting algorithm*, which splits the graph into its dendritic and cyclic part
- *decomposition algorithm* which transforms the cyclic part of the graph into a 'quasi dendritic' one
- *orientation algorithm* which directs all the edges according to the results of the previous two algorithms

Each of these algorithm is presented by its pseudo-code and the list of its inputs and outputs.

1. Splitting algorithm

INPUT - incidence matrix in the form of a two-dimensional array

OUTPUT- C-matrix - incidence matrix of the cyclic part
CV, CE - lists of vertices and edges respectively from the cyclic part
DV, DE - lists of vertices and edges respectively from the dendritic part

PSEUDO-CODE

```
initialise c-matrix as a copy of incidence matrix

repeat until number of found vertices = 0
begin
        found vertex = 0
        for i:=1 to number of vertices
        begin
                if vertex degree = 1 then
                begin
                        increase found vertex
                        place vertex number into DV
                        place the number of incident edge into DE
                        remove this edge from C-matrix
                end
        end
end

for i:= 1 to number of vertices
        if vertex degree <> 0 then
        begin
                place i in CV list
        else
                remove this row from C-matrix
        end

for i:= 1 to number of edges
        if sum of all entries in this vertical <> 0 then
        begin
                place i in CE list
        else
                remove this row from C-matrix
        end
```

2. Decomposition algorithm

INPUT - C-matrix, the incidence matrix of the cyclic part of the graph in the form of a two-dimensional array

OUTPUT- NS-number of symbols
S - list of symbol vertices (array)
NP - list with numbers of dendritic paths corresponding to each symbol (array)
LP - length of each dendritic path (2-dim. array)
V - list of vertices per each symbol and path (3-dim array)
E - list of edges per each symbol and path (3-dim array)

PSEUDO-CODE

```
initialise WM as a copy of C-matrix

for i:=1 to number of rows do
                              calculate degree of each vertex
for i:=1 to number of rows do
begin
        if degree = highest degree then
        begin
                remove this vertex from WM
                NS := NS + 1
                vertex number place in S
        end
        for j:=1 to number of edges
        begin
                if edge degree = 1 then
                begin
                        increase number of paths for this symbol
                        place the edge number in E
                        increase LP belonging to this symbol and path
                        if incoming vertex has degree = 1 then
                        begin
                                vertex number removed from WM to V
                                increase LP
                                incident edge removed from WM to E
                        end
                end
        end
end
```

3. Orientation algorithm

INPUT - incidence matrix, DV, DE, V, E

OUTPUT- MM-matrix - oriented incidence matrix

PSEUDO-CODE

```
initialisation of MM-matrix as a copy of incidence matrix

for i:= 1 to number of dendritic vertices do
                                        MM[DV[i],DE[i]] = -1

for i:= 1 to number of symbols do
begin
        for j:=1 to number of paths  symbol i do
        begin
                MM[ [ S [ i ] , E [ i , j , 1 ] ] = -1

                if length of the path > 1 then
                        for k = 2 to length of the path
                                MM[ [ S [ i ] , E [ i , j , k ] ] = -1
        end
end
```

APPENDIX 4.3

Generalised solution algorithm

If unknowns (Q and h) are just named X_j irrespectively of their type but only connected with their position in the staggered grid, both equations, of continuity and of momentum, can be written in the same form:

$$A_j X_{j-1} + B_j X_j + C_j X_{j+1} = D_j$$

The only exceptions arise when the values of the water depth are given at nodes which are denoted by Y_j. Thus, the first and the last equation of any edge has a somewhat different form:

$$A_1 Y_{el} + B_2 X_1 + C_1 X_2 = D_1$$

$$A_{jj} X_{jj-1} + B_{jj} X_{jj} + C_{jj} Y_{e2} = D_{jj}$$

The generalised solution algorithm splits the solution of any network into two parts, namely the solution of the dendritic part and the solution of the looped part. Correspondingly, the explanation of the solution algorithm will be given in two stages.

Dendritic part - branched network algorithm

The dendritic part of a network is solved by the double-sweep algorithm whose recurrence relations for the elimination sweep are

$$X_j^{n+1} = E_j \cdot X_{j+1}^{n+1} + F_j$$

with the expressions for the coefficients being:

$$E_j = \frac{-C_j}{A_j E_{j-1} + B_j} \qquad \text{and} \qquad F_j = \frac{D_j - A_j F_{j-1}}{A_j E_{j-1} + B_j}$$

The elimination sweep along each edge starts either from an end-vertex, where a boundary condition is defined and initial values for coefficients E_0 and F_0 can be determined from the boundary condition, or from an internal vertex, where internal compatibility conditions are used to define initial values of E_0 and F_0 .

End-vertices with defined boundary condition

If a water depth boundary condition is given, then initialisation of the recurrence relation is simple and straightforward:

$$E_0 = 0 \qquad \text{and} \qquad F_0 = h_{given}$$

However, if a discharge boundary condition is given, an additional continuity equation is employed to relate the discharge at this vertex to the discharge at the first discretisation point along the edge:

$$\frac{X_1 - Q_{given}}{\Delta x} + \frac{b}{\Delta t}\left(Y^{n+1}_{vertex} - Y^{n}_{vertex} \right) = 0$$

From this equation, the expressions for the coefficients follow as:

$$E_0 = -\frac{\Delta t}{b\,\Delta x} \qquad \text{and} \qquad F_0 = \frac{\Delta t\, Q_{given}}{b\,\Delta x} + Y^{n}_{vertex}$$

Internal vertices

At internal vertices the continuity equations are used as the internal conditions. The equivalences of the water levels are also internal conditions but these were already used when unknowns were assigned to the points, while the water level at the vertex (Y_{vertex}) belongs to all incident edges.

The continuity equation at a vertex has the form:

$$\sum Q_e + q = 0$$

with Q_e being the discharge at the very beginning of an edge (not at the first discretisation point along the edge), with e and q being the external discharges.

In order to get these values of discharges at the beginnings of the edges, continuity equations were employed in the same way as in the case of an end vertex. All the incident edges are divided into incoming ones and only one outgoing one (form the point of view of elimination, and not of flow). The additional continuity equation for each of the incoming edges gives:

$$\frac{Q_{in} - X_{jj}^{n+1}}{\Delta x} + \frac{b}{\Delta t}\left(Y^{n+1} - Y^{n}\right) = 0$$

while the last recurrence relation along that edge gives:

$$X_{jj}^{n+1} = E_{jj} \cdot Y^{n+1} + F_{jj}$$

After combining these two expressions we get:

$$Q_{in} = \left(E_{jj} - \frac{b\,\Delta x}{\Delta t}\right) Y^{n+1} + F_{jj} + \frac{b\,\Delta x}{\Delta t} Y^{n}$$

The continuity equation for the considered vertex (taking into account the conventions for signs) gives:

$$Q_{out} = \sum Q_{in} + q = 0$$

The additional continuity equation for the outgoing edge is:

$$\frac{X_1^{n+1} - Q_{out}}{\Delta x} + \frac{b}{\Delta t}\left(Y^{n+1} - Y^n\right) = 0$$

When for unknown Q_{out} and Q_{in} the expressions from the last two equations are substituted and the whole equation is rewritten to obtain the form:

$$Y^{n+1} = E_0 \cdot X_1^{n+1} + F_0$$

we get expressions for initial coefficients E_0 and F_o

$$E_0 = -\frac{1}{-\sum^{d-1} E_{jj} + \sum^{d} \frac{b\,\Delta x}{\Delta t}}$$

and

$$F_0 = \frac{\sum^{d-1} F_{jj} + \sum^{d} \frac{b\,\Delta x}{\Delta t} Y^n + q}{-\sum^{d-1} E_{jj} + \sum^{d} \frac{b\,\Delta x}{\Delta t}}$$

End vertex

At the end vertex of a dendritic graph, two different boundary conditions are possible. The boundary condition water level, given as a function of time, is directly substituted in the last recurrence relation as Y^{n+1} and the substitution sweep is initiated.

In the case of a discharge boundary condition, an additional continuity equation is again employed:

$$\frac{-Q_{given} - E_{jj}Y^{n+1} - F_{jj}}{\Delta x} + \frac{b}{\Delta t}\left(Y^{n+1} - Y^n\right) = 0$$

which after rearranging gives:

$$Y^{n+1} = \frac{Q_{given} + F_{jj} + \dfrac{b\,\Delta x}{\Delta t} Y^{n}}{\dfrac{b\,\Delta x}{\Delta t} - E_{jj}}$$

After this, the substitution sweep is made following the reverse order to that of the elimination sweep.

No end vertex

If there is no end vertex there is a whole cyclic part of the network which has to be solved by a different algorithm. Once that is solved, the values of water levels in the vertices which are positioned at the connections to the dendritic part (or parts) of the graph are directly substituted into the last recurrence relations of the corresponding edges and the substitution sweep starts from there.

Looped part - global elimination algorithm

The elimination sweep

In the elimination sweep, the recurrence relations along the edges have the form:

$$X_j = E_j \cdot X_{j+1} + F_j + \sum_{i=1}^{k} G_{i,j} \cdot Y_{v_i}$$

with: j - index along the edge

i - index through the *S*-array with the symbol-vertices' numbers of the symbol-vertices corresponding to this *k*-subset

k - index number of the subset and at the same time the part of the index number of the corresponding symbols

The expressions for the coefficients are:

$$E_j = -\frac{C_j}{A_j \cdot E_{j-1} + B_j} \quad , \qquad F_j = \frac{D_j - A_j \cdot F_{j-1}}{A_j \cdot E_{j-1} + B_j} \quad ,$$

$$G_{i,j} = -\frac{A_j \cdot G_{i,j-1}}{A_j \cdot E_{j-1} + B_j}$$

The initial values of the coefficients for the first edge in a path are:

$$E(e,0) = 0 \quad , \qquad F(e,0) = 0 \qquad \text{and} \qquad G(i,e,0) = 1$$

while for the other edges in the same path they are:

$$E(e,0) = VE(v) \quad , \qquad F(e,0) = VF(v) \qquad \text{and} \qquad G(i,e,0) = VG(i,v)$$

with v being the number of the preceding vertex (from the V-array) and VE, VF and VG being the coefficients of the continuity equation at that vertex rewritten in the form of the recurrence relations along the edges.

The continuity equations at vertices are rewritten in the following form:

$$Y_v = VE_v \cdot X_1 + VF_v + \sum_{i=1}^{k} VG_{i,v} \cdot Y_{v_i}$$

with: v - index number of the vertex
e - index number of the following edge (element from the E-array)
i - index through the S-array with the symbol-vertices' numbers
k - index number of the subset and at the same time number of the corresponding symbols

If a simplified version is taken, the discharges situated in the first discretisation point along each edge are taken to be equal to the actual discharges at vertices (no additional continuity equations being employed) and the expressions for the coefficients are:

$$VE_v = -\frac{1}{\sum_{ee=1}^{d_i-1} E_{ee_{jj}}} \quad , \qquad VF_v = -\frac{\sum_{ee=1}^{d_i-1} F_{ee_{jj}}}{\sum_{ee=1}^{d_i-1} E_{ee_{jj}}} \quad ,$$

$$VG_{v,i} = -\frac{\sum_{ee=1}^{d_i-1} G_{ee_{i,jj}}}{\sum_{ee=1}^{d_i-1} E_{ee_{jj}}}$$

with: d_i - degree of the vertex v
ee - index through the edges incident to the vertex v (except the following edge from the E-array), $ee=1$ *to* d_i-1

If however, the additional continuity equations are employed, the expressions for coefficients *VE*, *VF* and *VG* are different, resembling more those described above for the dendritic part of the network.

The intermediate sweep:

The recurrence relations along the edges in the intermediate sweep have the form:

$$X_j = P_j + \sum_{i=1}^{k} R_{i,j} \cdot Y_{v_i} \qquad \text{(A.4.3.1)}$$

with: e - index number of the edge (element from the E-array)
j - index along the edge
i - index through the S-array with numbers of decomposed vertices
k - index number of the subset and at the same time the number of the corresponding symbol

The expressions for the coefficients are:

$$P_j = E_j \cdot P_{j+1} + F_j \quad , \qquad R_{i,j} = E_j \cdot R_{i,j+1} + G_{i,j}$$

The last edge in a path is incident either to a vertex belonging to one of the following k-subsets or to a vertex in which a symbol is defined. Concerning the initial values of the coefficients, they can in both cases be expressed through the following formulae:

$$P_{jj} = E_{jj} \cdot VP_v + F_{jj} \ , \qquad R_{i,jj} = E_{jj} \cdot VR_{i,v} + G_{i,jj}$$

In the case that the final vertex of a path is an element of another k-subset, the coefficients $VP(v)$ and $VR(i,v)$ are as defined previously. In the other case they are:

$$VP_v = 0 \ , \qquad VR_{i,v} = 0 \ , i = 1,..,k-1 \quad \text{and} \qquad VR_{k,v} = 1$$

For other edges in the same path, they are:

$$P_j = E_j \cdot VP_v + F_j \ , \qquad R_{i,j} = E_j \cdot VR_{i,v} + G_{i,j}$$

with v being the number of the following vertex (from the V-array) and VP and VR being the coefficients of the continuity equation at that vertex rewritten in the form of the recurrence relations along the edges in the intermediate sweep.

The continuity equations at the vertices are rewritten in the following form:

$$Y_v = VP_v + \sum_{i=1}^{k} VR_{i,v} \cdot Y_{v_i}$$

The expressions for the coefficients are:

$$VP_v = VE_v \cdot P_1 + VF_v \ , \qquad VR_{i,v} = VE_v \cdot R_{i,1} + VG_{i,v}$$

with: v - index number of the vertex
e - index number of the following edge (element from the E-array)
i - index through the S-array with numbers of decomposed vertices
k - index number of the subset and at the same time the number of corresponding symbols

The closing equations

Upon completion of the intermediate sweep for one k-subset, a closing equation for the kth symbol is needed. The continuity equation at the decomposed vertex, where the kth symbol is defined, is used. As all the incident edges have already been traversed by the intermediate sweep, all the discharges involved in the continuity equation are expressed as functions only of k symbols. Hence, the kth symbol can easily be expressed as a function of the remaining k-1 symbols.

$$Y_{v_k} = VP_{v_k} + \sum_{i=1}^{k-1} VR_{v_k,i} \cdot Y_{v_i} \qquad \text{(A.4.3.2)}$$

with: $v(k)$ - index number of the symbol-vertex where the kth symbol is defined
i - index through the S-array with numbers of decomposed vertices where the symbols corresponding to this k-subset are defined
k - index number of the subset and the symbol for which this is the closing equation

The expressions for the coefficients are:

$$VP_{v_k} = -\frac{\sum_{ee=1}^{d_i} P_{ee_1}}{\sum_{ee=1}^{d_i} R_{ee_{k},1}} \, , \qquad VR_{v_k,i} = -\frac{\sum_{ee=1}^{d_i} R_{ee_{i},1}}{\sum_{ee=1}^{d_i} R_{ee_{k},1}}$$

with: d_i - degree of the kth symbol-vertex $(v(k))$
ee - index through the edges incident to the kth symbol-vertex $(v(k))$

The closing equation for the first symbol, $Y(S(1))$, will have only one term so that this will itself be the solution for the first symbol. The substitution sweep which will then follow will include all the edges and vertices and give solutions for all the unknowns.

The substitution sweeps

After a closing equation for a symbol, a substitution sweep through all k-subsets corresponding to this vertex is performed. During this sweep, recurrence relations for all the unknowns maintain the same form as before. The only differences are in the correction of the coefficients and the reduction of the summation index k by one. For the water levels at the vertices, the coefficients of equation (A.4.3.2) are corrected in the following way:

$$VP_v = VP_v + VR_{v,k} \cdot VP_{v_k} \ , \qquad VR_{v,i} = VR_{v,i} + VR_{v,k} \cdot VR_{v_k,i}$$

In a similar manner, the coefficients of equation (A.4.3.1) for unknowns along the edges are corrected as:

$$P_j = P_j + R_{j,k} \cdot VP_{v_k} \ , \qquad R_j = R_{j,i} + R_{j,k} \cdot VR_{i,v_k}$$

Upon completion of the substitution sweep corresponding to the kth symbol, the procedure continues with the intermediate sweep corresponding to the k-1th symbol.

APPENDIX 4.4

Algorithm for direct solution of systems of linear algebraic equations characterised by nearly-banded matrices

Introduction

Let us consider the system of linear equations

$$A x = b \tag{A.4.4.1}$$

with A being a square matrix of order $n \times n$ and x and b being column vectors of order n.

Let us then characterise a 'nearly-banded' matrix for a square matrix of order $n \times n$ as a matrix that is banded except for m of its rows, such that $m \ll n$, and even these m rows are not very full.

The necessary conditions for a nearly-banded matrix A to be solvable by this method are as follows:

- that the matrix A is reducible to a *doubly bounded block diagonal form* (*DBBDF*). This form of matrix can be represented in block form as:

$$\begin{bmatrix} A_{11} & & & & & A_{1N} \\ & A_{22} & & & & A_{2N} \\ & & \cdot & & & \cdot \\ & & & A_{ii} & & A_{iN} \\ & & & & \cdot & \cdot \\ & & & & A_{N-1N-1} & A_{N-1N} \\ A_{N1} & A_{N2} \;\cdot & A_{Ni} & \cdot & A_{NN-1} & A_{NN} \end{bmatrix}$$

where non-zero entries appear only in these blocks and the rest of the matrix is filled with zeros.

- that all the diagonal blocks A_{ii}, $i=1..N\text{-}1$, are themselves banded
- that the block A_{NN} is a diagonal matrix or it is equal to zero
- that all blocks $A_{N\,i}$ have only two non-zero entries, one in the first column and the other in the last column, and these two entries are not in the same row
- that all blocks $A_{i\,N}$ are transposes of $A_{N\,i}$, $A_{i\,N} = A_{N\,i}^T$.

If matrix A fulfils all these conditions than a matrix C can be formed from a set of column vectors with each of these being formed from a block $A_{N\,i}$ by adding up all the entries in each row. Each element of the matrix C is then

$$c_{ij} = \sum_{k=1}^{n_j} a_{ik} \tag{A.4.4.2}$$

with n_j being the number of columns of the block $A_{N\,j}$ and $a_{i\,k}$ being the element in the ith row and k^{th} column of the block $A_{N\,j}$. The matrix C formed in such a way is considered as an incidence matrix of a graph. In the rest of this work, this graph will be called a *network graph*. The solution algorithm relies extensively on this graph. Each group of blocks $[..A_{i\,i}....A_{i\,N}]$ corresponds to an edge of this graph, while the set of blocks $[A_{1\,N}\ A_{2\,N} \ldots\ A_{N\,N}]$ corresponds to the vertices of the graph.

Such systems of equations are usually associated with finite difference solutions of flow in hydrodynamic networks where each channel (or pipe) of the network is discretised into a considerable number of points. However, although this algorithm was developed for such network problems, it may be applied to the solution of other systems of linear equations provided that these fulfil the above requirements.

If the origin of the system of equations is a network problem then this graph coincides with the original network graph. Then groups of blocks $[..A_{i\,i}....A_{i\,N}]$ originate from the approximation of the conservation equations along the channels or pipes of the network while the set of blocks $[A_{1\,N}\ A_{2\,N} \ldots\ A_{N\,N}]$ corresponds to the continuity equations in the nodes of the network. In fact, in the numerical algorithms for solutions of network problems, the whole matrix A is never assembled. The blocks $[..A_{i\,i}....A_{i\,N}]$ are, together with the right hand side, efficiently stored in the so called *diagonal storage form* (Pissanetsky, 1984, p.13) that is particularly suitable for banded matrices. The blocks $[A_{1N}\ A_{2N} \ldots\ A_{NN}]$ are stored in the condensed form of the incidence matrix of the network graph (which is equivalent to matrix C).

Let us consider a network represented by its network graph in Fig.A.4.4.1 as an example suited to an explanation of the global elimination method.

A common practice in the field of direct methods for matrix solution is to represent matrices by graphs with the number of vertices being equal to the order of the matrix and the number of edges of the graph corresponding to the non-zero entries of the matrix. A graph of this type will henceforth be called a *matrix graph* in order to distinguish it from a network graph.

If each edge of the network from Fig.A.4.4.1 is discretised in only four points, the matrix graph takes the form schematised in Fig.A.4.4.2. Note that the usual number of discretisation points along edges is much higher, but for the sake of this example four points will suffice.

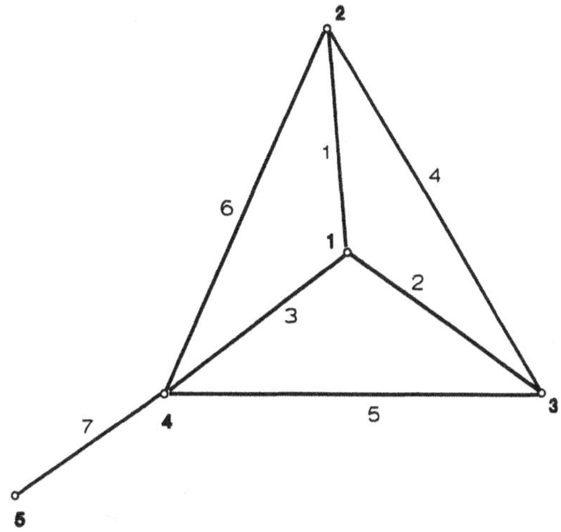

Fig.A.4.4.1

By the application of a *one way dissection* (George, 1980) or a *nested dissection* (George, 1973) or by *node tearing* (Duff *et al*, 1986, pp.254-261) such matrix graphs can then be split into sets of disjoint vertices by a *separator set*. This splitting results in a partitioning of the matrix in *DBBD* form and facilitates the solution of the system by any suitable economical method (i.e. Block Gauss Elimination, as described by Duff *et al*, 1986, pp.161-163). The choice of the separator set does influence the size of the block A_{NN} and consequently it does change the computational effort involved in the solution. The smaller separator set reduces drastically the computational effort (Duff *et al*, 1986, p.258). Hence, a successful dissection or node tearing is one which minimises the separator set and still maintains the separation of the other vertices.

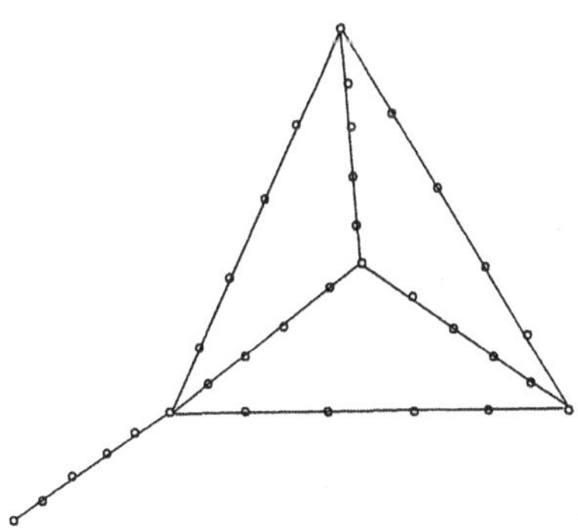

Fig.A.4.4.2

When the whole matrix A is represented by its matrix graph (as shown in Fig.A.4.4.2) then two types of vertices can be distinguished. There are those corresponding to the discretisation points along the channels and those corresponding to the actual vertices of the network graph. The most natural separator set for the dissection of such a graph is then the set of original vertices of the network graph. Only after being partitioned by the nested dissection does the matrix A obtain the *DBBD* form which was analysed earlier.

For the example network from Fig.A.4.4.1 (with only three discretisation points per edge) the *DBBD* form of the system matrix is given in equation (A.4.4.3).

The rows of the matrix are numbered in order to make it easier to follow later transformations of the matrix. The equations 1, 2 and 3 belong to the discretisation points situated along the edge 1, equations 4, 5 and 6 to those situated along the edge 2 and so on till arriving at equation 21. The equations 22 till 26 belong to the vertices of the network graph from 1 to 5 in the same order.

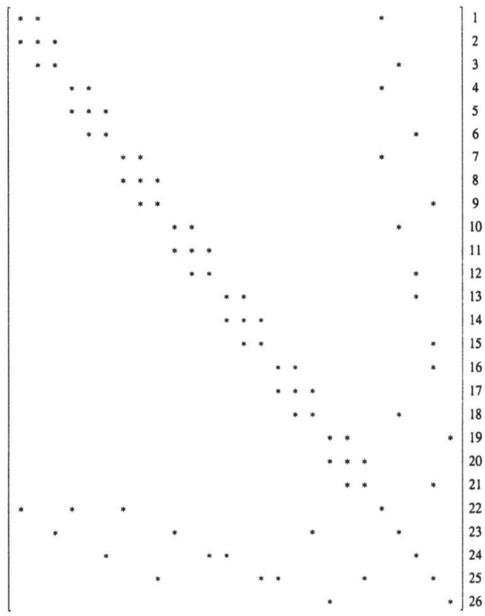

(A.4.4.3)

The classical methods for the solution of flows in looped channel networks uses this form of matrix A. Due to the significant number of nodes involved in some networks, the size of the separator set is also significant and consequently the size of the block A_{NN} is usually the bottle-neck in this type of algorithms. However, in the example presented here the number of vertices is small, which makes the size of the block A_{NN} rather insignificant. What is important is to notice that this number depends only on the number of vertices and not at all on the complexity of the network in terms of its degree of connectivity. Therefore, a tree-like network with n vertices, solved by the classical looped network algorithm, would also, completely unnecessarily, have a block A_{NN} of degree n.

The global elimination algorithm was developed with the aim of reducing the computational effort in such cases. The idea behind this method is to use the Gauss elimination, which is more precisely in its form for the solution of banded matrices called the double-sweep algorithm (Abbott, 1979, pp.166-177; Volkov, 1986, pp.155-160), and at the same time to try to minimise the fill-in originating from this elimination. The other methods such as those of the *minimal degree algorithm* (Tinney, 1969) are based on the same idea, but they apply it locally; e.g. the next pivot is chosen from the active sub-matrix on the basis of the minimal degree of the corresponding vertex in the graph. The classical algorithm for the solution of flow in looped channel networks follows a similar criterion. The equations originating from the discretisation points along the edges are eliminated first as they always have a minimal degree in comparison with the vertices of the network graph. However, once the problem is reduced to the solution of the block A_{NN}, different direct and indirect methods are normally applied.

The idea of the global elimination method is to use the topology of the network graph (which, we repeat, is not the same as that of the matrix graph) to define the 'strategy' of the elimination procedure. It is very well known that Gauss elimination applied to the matrix whose graph is a *tree* (*dendritic* graph) does not introduce any fill-in (Parter, 1961). Thus, in order to apply Gauss elimination with no fill-in, it is not necessary to have a completely separated graph, but it is sufficient that the graph is a tree. Following this reasoning, a method is developed which limits the number of elements in the separator set just to the number of vertices necessary to render the remaining parts of the graph tree-like. This means that as many nodes are being torn as are necessary to make the rest of the network graph 'quasi dendritic'.

As applied to our example graph from Fig.A.4.4.1, this node tearing or *decomposition* would provide the situation schematised in Fig.A.4.4.3. In this example it was necessary to decompose two vertices (number 1 and number 2) and then the rest of the graph became a tree. This means that the separator set has only these two vertices and not, as earlier, all five of them.

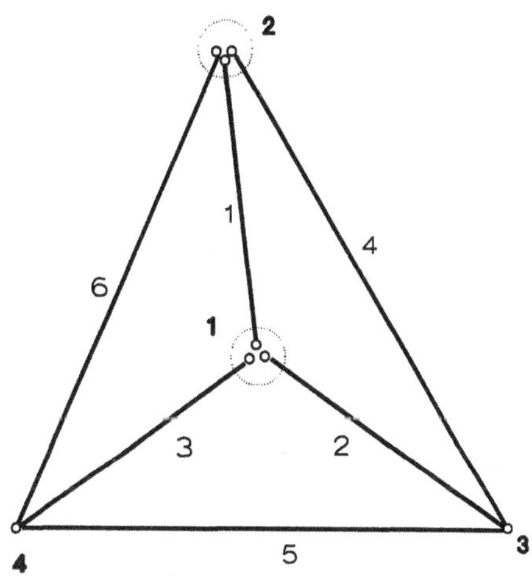

Fig.A.4.4.3

The fill-in that is generated in the global elimination method is limited at the columns corresponding to the torn vertices (that is, the vertices obtained from the separator set). The unknowns associated with these vertices are in the global elimination method simply called *symbols*. This name comes from the other form of writing the system of equations where the unknown associated with the diagonal element is expressed as a linear combination of other unknowns. In these equations, values of the unknowns at torn vertices then necessarily appear in symbolic form.

The global elimination algorithm can also be related to a method proposed by Wang (1981) for the solution of banded matrices using parallel processors. The difference in the present case is that the global elimination algorithm is applied to the solution of only nearly-banded matrices, and therefore the choice of the tearing nodes is governed by the topography of the network and not by the number of the intermediate vertices as the case in the work by Wang.

The reduced number of vertices in the separator set reduces the size of all the blocks $A_{N\,i}$, including the block $A_{N\,N}$, which reduces the required computational effort. However, the equations corresponding to the vertices of the network graph, which are now no longer elements of the separator set, are placed between the blocks $A_{i\,i}$ which they were separating in the previous partitioning. As a consequence, the new partitioned matrix no longer has a perfect *DBBD* form. In the following equation (A.4.4.4) the form of the matrix from our example is presented.

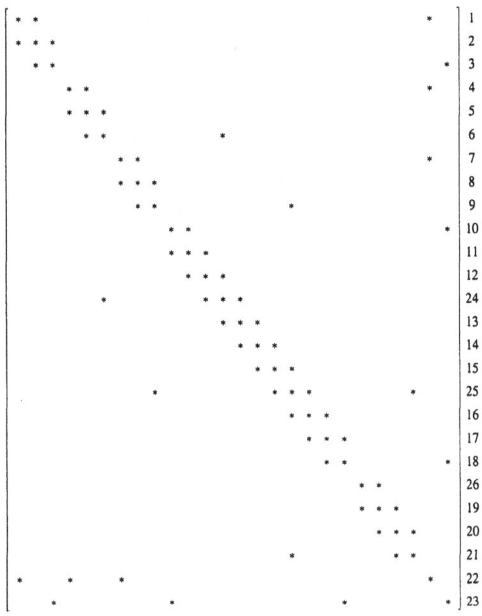

(A.4.4.4)

Due to the proper ordering of elimination, all the 'new' entries arising from out of the *DBBD* structure are easily eliminated and do not cause any additional fill-in. Thus, all the fill in will still be confined to the blocks $A_{i\,N}$. However, before the actual global elimination algorithm is explained, the definition of the sequence, which is the backbone of the global elimination algorithm, should be described.

Definition of the elimination sequence

A search procedure is performed on the incidence matrix to define the type of algorithm that will be applied and the order in which it will be executed. The search procedure consists of two parts, with the first searching for the *dendritic* (tree-like) part(s) of the graph and defining the order of elimination within these, and the second part defining the order of elimination among the edges belonging to the *cyclic* (looped) part of the graph. Formally:

First part of the search procedure

- search for the vertex with degree equal to one
- when found, place its number into the corresponding list and do the same with the only incident edge
- remove corresponding entries from the incidence matrix
- repeat this procedure until no such vertex is found in the pass through the whole incidence matrix.

Second part of the search procedure

- search for the vertex with maximum degree
- place its number in the list of symbols and remove all entries corresponding to it from the incidence matrix
- search for an edge which is incident to this vertex (a column with a single entry)
- when such an edge is found, place it into the list and remove it from the incidence matrix
- check the degree of the vertex incident to this edge; if equal to one, remove the vertex and the incident edge from the incidence matrix and place them into their corresponding lists
- repeat the last step as long as only vertices with a degree equal to one are encountered
- search for another edge incident to the originally removed vertex (a column with a single entry) and, when found, repeat the last three steps
- if no edges of that type are encountered and the incidence matrix is not empty, chose another vertex with the highest degree and place it into the list of symbols and repeat the last seven steps
- repeat this procedure until the incidence matrix is empty.

Upon application of this algorithm on the network from Fig.A.4.4.1. the obtained lists are:

dendritic edges	DE	=	{7}
dendritic vertices	DV	=	{5}
number of symbols	NS	=	2
list of symbol vertices	S		{1,2}
number of dendritic paths per each symbol	NP	=	{3,1}
length of each dendritic path	LP	=	{{1,1,1},{3}}
edges involved in each dendritic path	E	=	{{1},{2},{3},{4,5,6}}
vertices involved in each dendritic path	V	=	{∅,∅,∅,{3,4}}

The Algorithm

The algorithm consists of two sub-algorithms, corresponding to the solution of the dendritic part and the solution of the cyclic part. The explanation of each of the steps of the algorithm will be followed by the form that the transformations of matrix A would undergo after this step was applied.

Dendritic part

The dendritic part of the graph is defined by the lists of dendritic edges and vertices produced in the first part of the search procedure. To each edge from this list there corresponds a system of equations of the form

$$[\ldots A_{i\,i} \ldots\ldots A_{i\,N}]\, x = b_i$$

If reduced to only non-zero entries, this can be written as

$$[A_{i\,i} \quad A_{i\,N}]\, x_i = b_i$$

where x_i is a part of the unknown vector with only the unknowns from this edge.

The solution is obtained in two steps, just as in the classical double-sweep algorithm, these being again an elimination sweep and a substitution sweep. The elimination is performed according to the list of dendritic edges and the substitution sweep in the reverse order of the same list. In our example there is one dendritic edge (number 7) so the elimination sweep along it, starting from the end vertex (number 5), is performed first. The second part of the algorithm, being the substitution sweep, is performed only upon completion of the algorithm for the solution of the cyclic part of the graph.

Cyclic part

The elimination procedure on the cyclic part of the graph does introduce some fill-in. However, its extent is limited and well controllable. All the equations belonging to an edge listed after a definition of a symbol will have fill-in at the place of that symbol and at all the places corresponding to the already defined symbols there will be no fill-in at the columns corresponding to the symbols defined later in the search procedure.

All the additional fill-in is located in the block $A_{i\,N}$. The elimination within one edge is done according to the elimination procedure from the double-sweep corrected for the fill-in at defined symbols. The order of edges to be passed by the elimination sweep is defined by the list of cyclic edges defined in the second part of the search procedure. The elimination sweep is finished when all the edges from this list are exhausted. The form of the matrix A at the moment of initiation of the elimination sweep is shown in equation (A.4.4.5).

The intermediate sweep is performed according to the list of cyclic edges in the reverse order. It is similar to the substitution sweep in the classical double-sweep algorithm, except for the fill-in introduced in the elimination sweep. The intermediate sweep does not introduce any new fill-in.

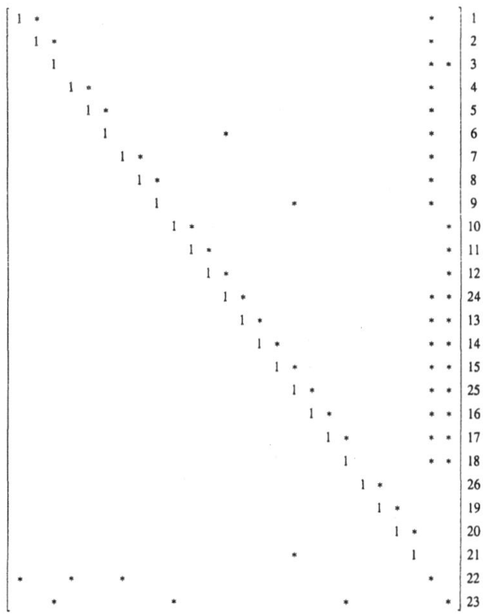

(A.4.4.5)

When the intermediate sweep along all the edges 'belonging' to the second symbol (edges 4, 5 and 6) is finished, the matrix A has the form shown in equation (A.4.4.6).

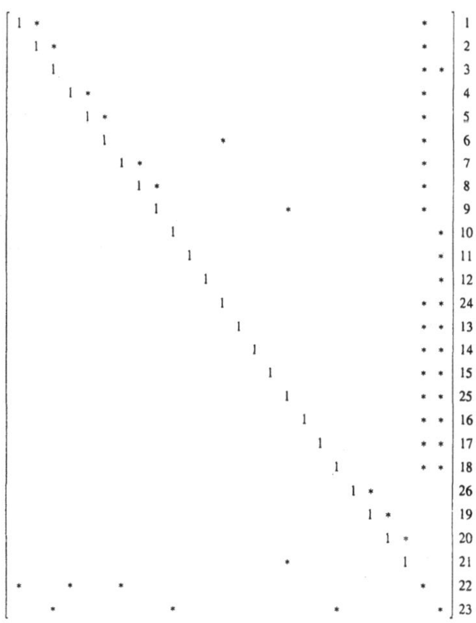

(A.4.4.6)

When the intermediate sweep for all the edges defined upon one symbol is finished, the corresponding equation from the matrix $[A_{N1}......A_{Ni}.....A_{NN}]$ is employed in order to express the value of that symbol as a function of all the previous symbols only. Substitution of this expression in all the equations having a fill-in at the place of that symbol reduces the amount of fill-in in these equations by one.

In our example, equation-23 (in the last row of the matrix) is the closing equation for the second symbol. All the non-zero entries of this equation, except these in the block A_{NN}, can be eliminated by equation-3, equation-10 and equation-18, and the second symbol can be expressed as a function of only the first symbol. The form of the matrix A upon the substitution sweep is given in equation(A.4.4.7).

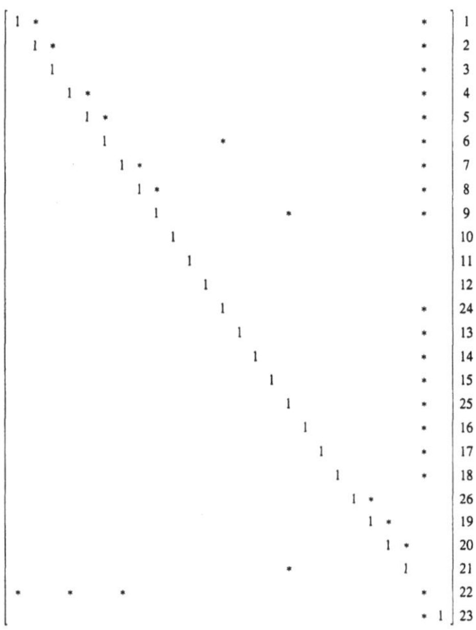

(A.4.4.7)

The procedure continues with the intermediate sweep for all the edges of the other symbol (that is, the one defined just before that which was just eliminated). The procedure is repeated until the value of symbol number one becomes known.

The form of the matrix A obtained after the intermediate sweep is performed on the edges belonging to the first symbol (edges 1,2 and 3) is given in equation(A.4.4.8). Then equation-22 is used as a closing equation for symbol one and this provides the value of this symbol.

After this, the substitution sweep throughout all the edges of the cyclic part of the graph is performed. With this the cyclic part of the graph is solved and the substitution sweep on the dendritic part of the graph can now be performed.

$$
\left[\begin{array}{cccccccccccccccccccccccccc}
1 &&&&& &&&&& &&&&& &&&&& &&&& * & \\
& 1 &&&&& &&&&& &&&&& &&&&& &&& * & \\
&& 1 &&&&& &&&&& &&&&& &&&&& && * & \\
&&& 1 &&&&& &&&&& &&&&& &&&&& & * & \\
&&&& 1 &&&&& &&&&& &&&&& &&&&& * & \\
&&&&& 1 &&&&& &&&&& &&&&& &&&& * & \\
&&&&& & 1 &&&&& &&&&& &&&&& &&& * & \\
&&&&& && 1 &&&&& &&&&& &&&&& && * & \\
&&&&& &&& 1 &&&&& &&&&& &&&&& & * & \\
&&&&& &&&& 1 &&&&& &&&&& &&&&& & \\
&&&&& &&&&& 1 &&&&& &&&&& &&&&& \\
&&&&& &&&&& & 1 &&&&& &&&&& &&&& \\
&&&&& &&&&& && 1 &&&&& &&&&& && * & \\
&&&&& &&&&& &&& 1 &&&&& &&&&& & * & \\
&&&&& &&&&& &&&& 1 &&&&& &&&&& * & \\
&&&&& &&&&& &&&&& 1 &&&&& &&&& * & \\
&&&&& &&&&& &&&&& & 1 &&&&& &&& * & \\
&&&&& &&&&& &&&&& && 1 &&&&& && * & \\
&&&&& &&&&& &&&&& &&& 1 &&&&& & * & \\
&&&&& &&&&& &&&&& &&&& 1 &&&&& * & \\
&&&&& &&&&& &&&&& &&&&& 1 & * &&&& \\
&&&&& &&&&& &&&&& &&&&& & 1 & * &&& \\
&&&&& &&&&& &&&&& &&&&& && 1 & * && \\
&&&&& &&&&& &&&&& & * &&&&& && 1 && \\
* &&& * &&& * &&&&& &&&&& &&&&& &&&& * & \\
&&&&& &&&&& &&&&& &&&&& &&&& * & 1
\end{array}\right]
\begin{array}{l}
1\\ 2\\ 3\\ 4\\ 5\\ 6\\ 7\\ 8\\ 9\\ 10\\ 11\\ 12\\ 24\\ 13\\ 14\\ 15\\ 25\\ 16\\ 17\\ 18\\ 26\\ 19\\ 20\\ 21\\ 22\\ 23
\end{array}
\qquad (A.4.4.8)
$$

Comment

It is interesting to observe that the solution of the dendritic part of the graph, being the double sweep algorithm guided by the list of edges and vertices to be followed, can be seen as equivalent to the algorithm for the cyclic part of the graph. It only needs to have the number of the symbols being equal to zero and consequently the set of all symbols to be an empty set in the case of a branched network (that is, a tree-like graph) the global elimination algorithm reduces to the double-sweep algorithm.

In that respect, the global elimination algorithm can be viewed as a general method for the solution of banded and nearly banded matrices, where the algorithm for the solution of dendritic networks (branched algorithm) is just a trivial case of it with an empty set of symbols.

APPENDIX 5.1

The standard ADI algorithm used in Chapter 5

The alternated direction implicit (ADI) algorithm was applied to solve the shallow water equations:

$$\frac{\partial f}{\partial t} + \frac{\partial g_1}{\partial x_1} + \frac{\partial g_2}{\partial x_2} = 0 \qquad \text{(A.5.1.1)}$$

with:

$$f = \begin{bmatrix} h \\ u_1 h \\ u_2 h \end{bmatrix}, \qquad g_1 = \begin{bmatrix} u_1 h \\ \frac{gh^2}{2} \\ 0 \end{bmatrix}, \qquad g_2 = \begin{bmatrix} u_2 h \\ 0 \\ \frac{gh^2}{2} \end{bmatrix}$$

where:

h - water depth
u - velocity in x direction
v - velocity in y direction
g - acceleration due to gravity

The same set of equations can be rewritten in another form as:

$$\frac{\partial h}{\partial t} + \frac{\partial u_1 h}{\partial x_1} + \frac{\partial u_2 h}{\partial x_2} = 0$$

$$\frac{\partial u_1 h}{\partial t} + gh\frac{\partial h}{\partial x_1} = 0$$

$$\frac{\partial u_2 h}{\partial t} + gh\frac{\partial h}{\partial x_2} = 0$$

or

$$\frac{\partial h}{\partial t} + \bar{h}\left(\frac{\partial u_1}{\partial x_1} + \frac{\partial u_2}{\partial x_2}\right) = 0 \qquad \text{(A.5.1.2)}$$

$$\frac{\partial u_1}{\partial t} + g\frac{\partial h}{\partial x_1} = 0 \qquad \text{(A.5.1.3)}$$

$$\frac{\partial u_2}{\partial t} + g\frac{\partial h}{\partial x_2} = 0 \qquad \text{(A.5.1.4)}$$

The ADI algorithm splits the calculation of one time step into two parts. Each of these parts is associated with one of the main coordinate directions (x_1 and x_2). In each direction the equation of conservation of mass is solved together with the equation of conservation of momentum in that direction.

The domain is discretised according to the grid presented in Fig.A.5.1.1.

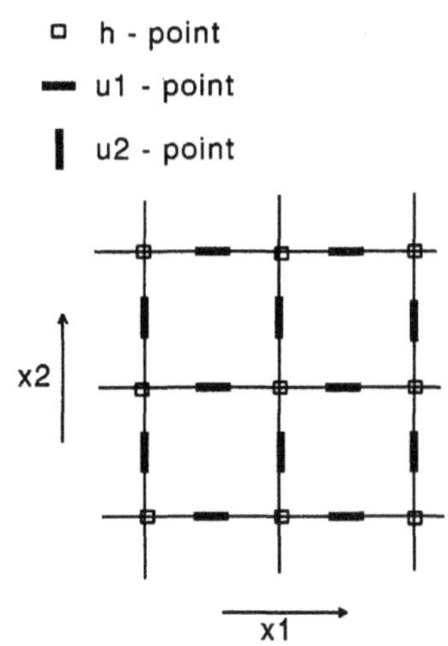

Fig.A.5.1.1

The equation of conservation of mass which is being solved in both directions ensures that a unique solution is obtained. Therefore, two continuity equations applied at the same point share the time step between them; the term $\partial h/\partial t$ is in each of these equations approximated on a half time step: in the x_1 -direction, between the time step n and $n+1/2$ and in the direction x_2 between the time step $n+1/2$ and the time step $n+1$.

Each part of an ADI algorithm consists of a repetition of algorithms for the solution of unknowns along one line which are placed along all grid lines in that particular direction. Along one grid line only the water depths and their corresponding velocities are discretised implicitly, which necessitates that they be solved simultaneously. Thus, the algorithm for the solution of one line produces a system of linear equations characterised by a tri-diagonal matrix. For the solution of this system of linear equations, the double-sweep algorithm is used.

The boundary conditions supported by this model are the velocity and the water depth as functions of time. (A closed end is considered to have a velocity equal to zero.) If a water depth boundary condition is given, then the boundary of the domain is at an h point . If a velocity is given then the boundary of the domain is placed at a velocity point. This approach makes the introduction of the boundary conditions much simpler (Verboom *et* al, 1992).

The algorithm is presented in two parts, the first defining the finite difference approximation and the second the solution algorithm.

Finite difference approximations

Upon the introduction of finite difference approximations instead of partial derivatives in equations (A.5.1.2) up to (A.5.1.4), each equation can be written in the following form:

$$A_j z_{j-1}^{n+1} + B_j z_j^{n+1} + C_j z_{j+1}^{n+1} = D_j$$

where details about coefficients and dependant variables depend on the original equation.

Sweeps along the lines parallel with the x_1 direction

Equation of conservation of mass:

$$A_{j,k} u_{1_{j-1,k}}^{n+1} + B_{j,k} h_{j,k}^{n+1/2} + C_{j,k} u_{1_{j+1,k}}^{n+1} = D_{j,k}$$

with coefficients:

$$A_{j,k} = -\frac{\bar{h}\,\Delta t}{4\,\Delta x}\ , \qquad B_{j,k} = 1\ ,$$

$$C_{j,k} = \frac{\bar{h}\,\Delta t}{4\,\Delta x}$$

$$D_{j,k} = h_{j,k}^{n} - \frac{\bar{h}\,\Delta t}{4\,\Delta x}\left(u_{1_{j+1,k}}^{\ n} - u_{1_{j-1,k}}^{\ n}\right) - \frac{\bar{h}\,\Delta t}{2\,\Delta x}\left(u_{2_{j,k+1}}^{\ n} - u_{2_{j,k-1}}^{\ n}\right)$$

Equation of conservation of momentum:

$$A_{j,k} h_{j-1,k}^{n+1/2} + B_{j,k} u_{1_{j,k}}^{\ n+1} + C_{j,k} h_{j+1,k}^{n+1/2} = D_{j,k}$$

with coefficients:

$$A_{j,k} = -\frac{g\,\Delta t}{\Delta x}\ , \qquad B_{j,k} = 1\ ,$$

$$C_{j,k} = \frac{g\,\Delta t}{\Delta x}\ , \qquad D_{j,k} = u_{1_{j,k}}^{\ n}$$

Sweeps along the lines parallel with the x_2 *direction*

Equation of conservation of mass:

$$A_{j,k} u_{2_{j,k-1}}^{n+1} + B_{j,k} h_{j,k}^{n+1} + C_{j,k} u_{2_{j,k+1}}^{n+1} = D_{j,k}$$

with coefficients:

$$A_{j,k} = -\frac{\bar{h}\,\Delta t}{2\,\Delta x} , \qquad B_{j,k} = 1 ,$$

$$C_{j,k} = \frac{\bar{h}\,\Delta t}{2\,\Delta x}$$

$$D_{j,k} = h_{j,k}^{n+1/2} - \frac{\bar{h}\,\Delta t}{4\,\Delta x}\left(u_{1_{j+1,k}}^{n+1} - u_{1_{j-1,k}}^{n+1} + u_{1_{j+1,k}}^{n} - u_{1_{j-1,k}}^{n} \right)$$

Equation of conservation of momentum:

$$A_{j,k} h_{j,k-1}^{n+1} + B_{j,k} u_{2_{j,k}}^{n+1} + C_{j,k} h_{j,k+1}^{n+1} = D_{j,k}$$

with coefficients:

$$A_{j,k} = -\frac{g\,\Delta t}{2\,\Delta x} , \qquad B_{j,k} = 1 ,$$

$$C_{j,k} = \frac{g\,\Delta t}{2\,\Delta x} , \qquad D_{j,k} = u_{2_{j,k}}^{n} - \frac{g\,\Delta t}{2\,\Delta x}\left(h_{j,k+1}^{n} - h_{j,k-1}^{n} \right)$$

Solution algorithm

The solution algorithm is given in the form of the following pseudo-code:

```
loop on all time steps
begin
        loop on all x1-lines
        begin
                calculate coefficients
                values of the boundaries
                solution of this line by the double-sweep
        end
        loop on all x2-lines
        begin
                calculate coefficients
                values of the boundaries
                solution of this line by the double-sweep
        end
end
```

APPENDIX 5.2

ADI+ algorithm - the combination of the ADI and the new algorithm

The ADI+ algorithm is an extension of the ADI algorithm for the solution of the shallow water equations describing two-dimensional nearly-horizontal flow. It in fact consists of two ADI algorithms applied within one time step. Each of these algorithms is applied on a different discretisation grid. However, the *h* points of these two grids correspond while the main coordinate directions are rotated through 45° as schematised in Fig.A.5.2.1.

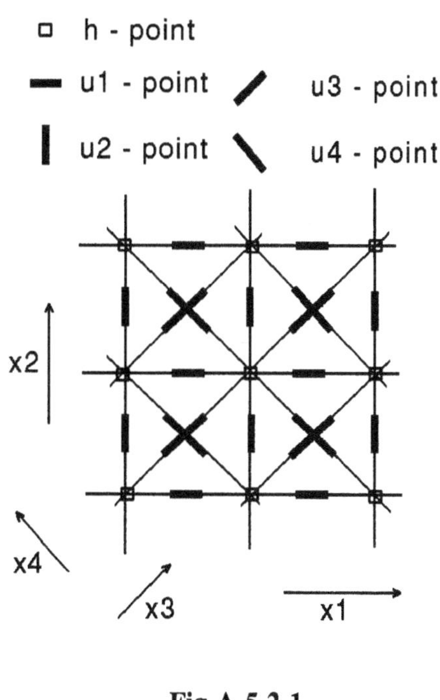

Fig.A.5.2.1

Although these two grids, x_1-x_2 and x_3-x_4, seem at the first sight similar, albeit with different discretisation steps, if observed separately it becomes clear that they differ in one special respect . In fact, the x_3-x_4 grid itself consists of two superimposed grids of the same type as the grid x_1-x_2 . This fact might lead to spurious oscillations of solutions, and accordingly a special attention is paid to it.

The shallow water equations can be written in the x_1-x_2 coordinate system as:

$$\frac{\partial f}{\partial t} + \frac{\partial g_1}{\partial x_1} + \frac{\partial g_2}{\partial x_2} = 0 \tag{A.5.2.1}$$

with:

$$f = \begin{bmatrix} h \\ u_1 h \\ u_2 h \end{bmatrix}, \quad g_1 = \begin{bmatrix} u_1 h \\ u_1^2 h + \dfrac{gh^2}{2} \\ u_1 u_2 h \end{bmatrix}, \quad g_2 = \begin{bmatrix} u_2 h \\ u_2 u_1 h \\ u_2^2 h + \dfrac{gh^2}{2} \end{bmatrix}$$

where:

x_1,x_2	- space coordinates	u_1	- velocity in x_1 direction
t	- time coordinate	u_2	- velocity in x_2 direction
h	- water depth	g	- acceleration due to gravity

Clearly, they can as well be written in the other grid x_3-x_4 as:

$$\frac{\partial f}{\partial t} + \frac{\partial g_3}{\partial x_3} + \frac{\partial g_4}{\partial x_4} = 0$$

with:

$$f = \begin{bmatrix} h \\ u_3 h \\ u_4 h \end{bmatrix}, \quad g_3 = \begin{bmatrix} u_3 h \\ u_3^2 h + \dfrac{gh^2}{2} \\ u_3 u_4 h \end{bmatrix}, \quad g_4 = \begin{bmatrix} u_4 h \\ u_4 u_3 h \\ u_4^2 h + \dfrac{gh^2}{2} \end{bmatrix}$$

and:

x_3,x_4 - space coordinates
u_3 - velocity in x_3 direction
u_4 - velocity in x_4 direction

For the purpose of the development of this new algorithm, only the simplest form of the shallow water equations is used, namely the form without the convective terms. In that form, the six equations (three for each coordinate system) are as follows:

in the x_1-x_2 coordinate system:

$$\frac{\partial h}{\partial t} + \bar{h} \left(\frac{\partial u_1}{\partial x_1} + \frac{\partial u_2}{\partial x_2} \right) = 0 \qquad \text{(A.5.2.2)}$$

$$\frac{\partial u_1}{\partial t} + g \frac{\partial h}{\partial x_1} = 0 \qquad \text{(A.5.2.3)}$$

$$\frac{\partial u_2}{\partial t} + g \frac{\partial h}{\partial x_2} = 0 \qquad \text{(A.5.2.4)}$$

in the x_3 - x_4 coordinate system:

$$\frac{\partial h}{\partial t} + \bar{h} \left(\frac{\partial u_3}{\partial x_3} + \frac{\partial u_4}{\partial x_4} \right) = 0 \qquad \text{(A.5.2.5)}$$

$$\frac{\partial u_3}{\partial t} + g \frac{\partial h}{\partial x_3} = 0 \qquad \text{(A.5.2.6)}$$

$$\frac{\partial u_4}{\partial t} + g \frac{\partial h}{\partial x_4} = 0 \qquad \text{(A.5.2.7)}$$

Solution algorithm

The ADI+ algorithm splits the calculation of one time step into four parts. The time step is basically divided into two parts, namely $w\Delta t$ and $(1-w)\Delta t$, and each of these parts is associated with one of the coordinate systems (x_1-x_2 and x_3-x_4 respectively). In each of these parts a 'standard' ADI algorithm is applied. This splits each of these parts of the time step further into two parts corresponding to the coordinate directions. In each of the directions the equation of conservation of mass is solved together with the equation of conservation of momentum in that direction.

The solution algorithm is presented in the form of the following pseudo-code:

```
loop on all time steps
begin
        calculation of weighting coefficient

        loop on all x1-lines
        begin
                solution of unknowns u1 at n+1 time level
                and h at n+w/2 time level
        end

        loop on all x2-lines
        begin
                solution of unknowns u2 at n+1 time level
                and h at n+w time level
        end

        loop on all x3-lines
        begin
                solution of unknowns u3 at n+1 time level
                and h at n+w+(1-w)/2 time level
        end

        loop on all x4-lines
        begin
                solution of unknowns u4 at n+1 time level
                and h at n+1 time level
        end
end
```

The solution along each line consists of calculations of the corresponding coefficients *A, B, C* and *D* following the finite difference approximations and the solution of the resulting system of linear simultaneous equations is provided by the double-sweep algorithm. For details of the solution algorithm, see Appendix 5.1, while the formulae for coefficients for all the directions follow shortly.

Finite difference approximations

The discretisation indices are *j, k, l* and *m* in directions x_1, x_2, x_3 and x_4 respectively. The space step in directions x_1 and x_2 is equal to Δx while in the directions x_3 and x_4 it is equal to Δs. The time discretisation step is Δt while the weighting coefficient *w* is calculated at each point of the domain from the results obtained at the previous time step. The weighting coefficient can only take values between 0 and 1. Upon the introduction of finite difference approximations instead of partial derivatives in equations (A.5.2.2) to (A.5.2.7), each equation can be written in the following form:

$$A_j z_{j-1}^{n+1} + B_j z_j^{n+1} + C_j z_{j+1}^{n+1} = D_j$$

where the details of the coefficients and dependant variables depend on the direction along which the equation is taken.

Sweeps along the lines parallel with the x_1 direction:

Equation of conservation of mass (A.5.2.2):

$$A_{j,k} u_{1_{j-1,k}}^{n+w} + B_{j,k} h_{j,k}^{n+w/2} + C_{j,k} u_{1_{j+1,k}}^{n+w} = D_{j,k}$$

with coefficients:

$$A_{j,k} = -\frac{w \bar{h} \Delta t}{4 \Delta x} \quad , \qquad B_{j,k} = 1 \quad ,$$

$$C_{j,k} = \frac{w \bar{h} \Delta t}{4 \Delta x} \quad ,$$

$$D_{j,k} = h_{j,k}^n - \frac{w \bar{h} \Delta t}{4 \Delta x} \left(u_{1_{j+1,k}}^n - u_{1_{j-1,k}}^n \right) - \frac{w \bar{h} \Delta t}{2 \Delta x} \left(u_{2_{j,k+1}}^n - u_{2_{j,k-1}}^n \right)$$

Equation of conservation of momentum in x_1 direction (A.5.2.3):

$$A_{j,k} h_{j-1,k}^{n+w/2} + B_{j,k} u_{1_{j,k}}^{n+w} + C_{j,k} h_{j+1,k}^{n+w/2} = D_{j,k}$$

with coefficients:

$$A_{j,k} = -\frac{w\,g\,\Delta t}{4\,\Delta x} \quad , \qquad B_{j,k} = 1 \quad ,$$

$$C_{j,k} = \frac{w\,g\,\Delta t}{4\,\Delta x} \quad , \qquad D_{j,k} = u_{1_{j,k}}^{n}$$

Sweeps along the lines parallel with the x_2 direction:

Equation of conservation of mass (A.5.2.2):

$$A_{j,k} u_{2_{j,k-1}}^{n+w} + B_{j,k} h_{j,k}^{n+w} + C_{j,k} u_{2_{j,k+1}}^{n+w} = D_{j,k}$$

with coefficients:

$$A_{j,k} = -\frac{w\,\bar{h}\,\Delta t}{2\,\Delta x} \quad , \qquad B_{j,k} = 1 \quad ,$$

$$C_{j,k} = \frac{w\,\bar{h}\,\Delta t}{2\,\Delta x} \quad ,$$

$$D_{j,k} = h_{j,k}^{n+w/2} - \frac{w\,\bar{h}\,\Delta t}{4\,\Delta x}\left(u_{1_{j+1,k}}^{n} - u_{1_{j-1,k}}^{n} + u_{1_{j+1,k}}^{n+w} - u_{1_{j-1,k-1}}^{n+w}\right)$$

Equation of conservation of momentum in x_2 direction (A.5.2.4):

$$A_{j,k} h_{j,k-1}^{n+w} + B_{j,k} u_{2_{j,k}}^{n+w} + C_{j,k} h_{j,k+1}^{n+w} = D_{j,k}$$

with coefficients:

$$A_{j,k} = -\frac{w\,g\,\Delta t}{2\,\Delta x} \quad , \qquad B_{j,k} = 1 \quad ,$$

$$C_{j,k} = \frac{w\,g\,\Delta t}{2\,\Delta x} \quad ,$$

$$D_{j,k} = u_{2_{j,k}}^{n} - \frac{w\,g\,\Delta t}{2\,\Delta x}\left(h_{j,k+1}^{n} - h_{j,k-1}^{n}\right)$$

Sweeps along the lines parallel with the x_3 direction:

Equation of conservation of mass (A.5.2.5):

$$A_{l,m} u_{1_{l-1,m}}^{n+1} + B_{l,m} h_{l,m}^{n+w+(1-w)/2} + C_{l,m} u_{1_{l+1,m}}^{n+1} = D_{l,m}$$

with coefficients:

$$A_{l,m} = -\frac{(1-w)\,\bar{h}\,\Delta t}{4\,\Delta x} \quad , \qquad B_{l,m} = 1 \quad ,$$

$$C_{l,m} = \frac{(1-w)\,\bar{h}\,\Delta t}{4\,\Delta x} \quad ,$$

$$D_{l,m} = h_{l,m}^{n} - \frac{(1-w)\,\bar{h}\,\Delta t}{4\,\Delta x}\left(u_{3_{l+1,m}}^{n} - u_{3_{l-1,m}}^{n}\right) - \frac{(1-w)\,\bar{h}\,\Delta t}{2\,\Delta x}\left(u_{4_{l,m+1}}^{n} - u_{4_{l,m-1}}^{n}\right)$$

Equation of conservation of momentum in x_3 direction (A.5.2.6):

$$A_{l,m}h_{l-1,m}^{n+w+(1-w)/2} + B_{l,m}u_{3_{l,m}}^{n+1} + C_{l,m}h_{l+1,m}^{n+w+(1-w)/2} = D_{l,m}$$

with coefficients:

$$A_{l,m} = -\frac{(1-w)\,g\,\Delta t}{\Delta s} \quad , \qquad B_{j,k} = 1 \quad ,$$

$$C_{l,m} = \frac{(1-w)\,g\,\Delta t}{\Delta s} \quad , \qquad D_{l,m} = u_{3_{l,m}}^{n}$$

Sweeps along the lines parallel with the x_4 direction:

Equation of conservation of mass (A.5.2.5):

$$A_{l,m}u_{4_{l,m-1}}^{n+1} + B_{l,m}h_{l,m}^{n+1} + C_{l,m}u_{4_{l,m+1}}^{n+1} = D_{l,m}$$

with coefficients:

$$A_{l,m} = -\frac{(1-w)\,\bar{h}\,\Delta t}{2\,\Delta s} \quad , \qquad B_{l,m} = 1 \quad ,$$

$$C_{l,m} = \frac{(1-w)\,\bar{h}\,\Delta t}{2\,\Delta s} \quad ,$$

$$D_{l,m} = h_{l,m}^{n+w+(1-w)/2} - \frac{(1-w)\,\bar{h}\,\Delta t}{4\,\Delta x}\left(u_{3_{l+1,m}}^{\,n} - u_{3_{l-1,m}}^{\,n} + u_{3_{l+1,m}}^{\,n+1} - u_{3_{l-1,m-1}}^{\,n+1}\right)$$

Equation of conservation of momentum in x_4 direction (A.5.2.7):

$$A_{l,m} h_{l,m-1}^{n+1} + B_{l,m} u_{4_{l,m}}^{\,n+1} + C_{l,m} h_{l,m+1}^{n+1} = D_{l,m}$$

with coefficients:

$$A_{j,k} = -\frac{(1-w)\,g\,\Delta t}{2\,\Delta s} \quad , \qquad B_{l,m} = 1 \quad ,$$

$$C_{l,m} = \frac{(1-w)\,g\,\Delta t}{2\,\Delta s} \quad ,$$

$$D_{l,m} = u_{4_{l,m}}^{\,n} - \frac{(1-w)\,g\,\Delta t}{2\,\Delta s}\left(h_{l,m+1}^{n+w+(1-w)/2} - h_{l,m-1}^{n+w+(1-w)/2}\right)$$

References

Abbott, M.B. (1966). *An Introduction to the Method of Characteristics*, Thames & Hudson, London, and American Elsevier, New York

Abbot, M.B., Damsgard, A. and Rodenhuis, G.S. (1973). *System 21 "Jupiter" - A design system for two-dimensional nearly-horizontal flows*, Journal of Hydraulic Research, Vol. 11, No.1, pp. 1-28

Abbott, M.B. (1976). *Computational Hydraulics: A Short Pathology*, Journal of Hydraulic Research, Vol. 14, No. 3, pp. 271-285

Abbott, M.B. (1979). *Computational Hydraulics: Elements of the Theory of Free Surface Flows*, Pitman Publishing Limited, London

Abbott, M.B., McCowan, A.D. and Warren, I.R. (1981). *Numerical modelling of free-surface flows that are two-dimensional in plane*, in Fischer, H.B. (editor), *Transport Models for Inland and Coastal Waters, Symposium on Predictive Ability*, Academic, New York

Abbott, M.B. and Basco, D.R. (1989). *Computational Fluid Dynamics : An Introduction for Engineers*, Longman, London.

Abbott, M.B., Havnø, K. and Lindberg, S. (1991). *The fourth generation of numerical modelling in hydraulics,* Journal of Hydraulic Research. Vol. 29, No. 5, pp. 581-600

Anderson, D.A. (1983) *Application of Adaptive Grids to transient Problems,* in Babuška, I., Chandra, J. and Flaherty, J.E. (Editors), *1. Workshop on adaptive computational methods for partial differential equations*, pp. 208-223, SIAM, Philadelphia

Ashby, W.R. (1973). *An Introduction to Cybernetics*, Chapman&Hall Ltd, London

Åström, K.J. and Wittenmark, B. (1989). *Adaptive Control,* Addison Wesley Publishing Company, USA

Atlas, I. and Stephenson, J.W. (1991). *A Two-Dimensional Adaptive Mesh Generation Method*, Journal of Computational Physics 94, pp. 201-224

Babuška, I., Chandra, J. and Flaherty, J.E. (Editors), (1983). *1. Workshop on adaptive computational methods for partial differential equations*, SIAM, Philadelphia

Babuška, I.(1989). *Adaptive Mathematical Modeling,* pp. 1-14, in Flaherty, J.E., Paslow, P.J., Shephard,M.S. and Vasilakis, J.D. (Editors), *Adaptive Methods for Partial Differential Equations*, SIAM, Philadelphia

Berger M.J. and Oliger, J. (1984). *Adaptive Mesh Refinement for hyperbolic Partial Differential Equations,* Journal of Computational Physics 53, pp. 484-512

Carré, B. (1979). *Graphs and Networks*, Clarendon, Oxford

Chakravarthy, S.R. and Szema, K.-Y. (1989). *Advances in finite difference techniques for computational fluid dynamics*, in Noor, A.K. and Oden J.T. (Editors), *State-of-the-art surveys on computational mechanics*, American Society of Mechanical Engineers, New York

Chaudry, M.H. (1993). *Open-Channel Flow,* Prentice-Hall, Inc., New Jersey

Chen, W. K. (1990). *Theory of Nets : Flow in Networks*, Wiley, New York

Cunge, J.A., Holly, F.M. and Verwey, A. (1980). *Practical Aspects of Computational River Hydraulics*, Pitman, London

Derrida, J. (1991). *Letter to a Japanese Friend*, in Kamuff, P. *A Derrida reader : Between the blinds*, Columbia University Press, USA

DHI, (1987-1992). *MIKE11 - Scientific documentation*, Danish Hydraulic Institute,Hørsholm

DHI, (1992). *MIKE21 - Scientific Background*, Danish Hydraulic Institute,Hørsholm

Duff, I.S., Erisman, A.M. and Reid, J.K. (1986). *Direct Methods for Sparse Matrices*, Claredon Press, Oxford

Garcia-Navarro, P. and Saviron, J.M. (1992). *McCormack's method for the numerical simulation of one dimensional discontinuous unsteady open channel flow*, Journal of Hydraulic Research, Vol. 30, No. 1, pp. 95-106

Garcia-Navarro, P. and Saviron, J.M. (1992). *Numerical simulation of unsteady flow at open channel junctions*, Journal of Hydraulic Research, Vol. 30, No. 5, pp. 595-610

Garcia-Navarro, P., Priestley A. and Alcrudo, F. (1994). *An implicit method for water flow modelling in channels and pipes*, Journal of Hydraulic Research, Vol. 32, No. 5, pp. 721-742

George, A. (1973). *Nested dissection of a regular finite-element mesh*, SIAM J. Numer. Anal. 10, pp. 345-363

George, A. (1980). *An automatic one-way dissection algorithm for irregular finite element problems*, SIAM J. Numer. Anal. 17, pp. 740-751

George, A. and Liu, J.W.H. (1989). *The evolution of the minimum degree ordering algorithm*, SIAM Review, Vol. 31. No. 1, pp. 1-19

Godunov, S.K. and Ryabenki, V.S. (1964). *Theory of Difference Schemes*, Nord Holland Publishing Company, Amsterdam

Harten, A. (1983). *Adaptive Numerical Methods for Hyperbolic Conservation Laws,* in Babuška, I., Chandra, J. and Flaherty, J.E. (Editors), *1. Workshop on adaptive computational methods for partial differential equations*, pp. 224-236, SIAM, Philadelphia

Havnø, K. and Brorsen, M. (1985). *Generalised mathematical modelling system for flood analysis and flood control design, Papers 2. International Conference on Hydraulics of Floods and Flood Control*, Cambridge, UK, BHRA, Stevenage

Isermann, R., Lachmann, K.-H. and Matko, D. (1992). *Adaptive Control Systems*, Prentice Hall International Ltd, UK

Klaassen, G.J. and Urk, A.van (1985). *Resistance to Flow of Flood Plains with Grasses and Hedges, 21st Congress, 1985, Melbourne, Australia.*

Kortteve, W. (1994). *Wave Attenuation by Using Reed for Bank Protection*, Delft Univ. of Technology, Report.

Kutija, V. (1993). *On the numerical modelling of supercritical flow*, Journal of Hydraulic Research, Vol. 31, No. 6, pp. 841-858

Kutija, V. (1995). *A generalised method for the solution of flows in networks,* Journal of Hydraulic Research, Vol. 33, No. 4, pp. 535-554

Kutija, V. and Hong, H.T.M. (1996). *A numerical model for assessing the additional resistance to flow introduced by flexible vegetation*, Journal of Hydraulic Research, Vol. 34, No. 1, pp. 99-114

Larsen, T., Frier, J.O. and Vestergraard, K. (1990). *Discharge/Stage Relation in Vegetated Danish Stream, International Conference on River Flood Hydraulics, 1990*

Lerner, A.Ya. (1972). *Fundamentals of Cybernetics*, Chapmann & Hall Ltd,London

Liggett, J.A. and Cunge, J.A. (1975). *Numerical methods of solution of the unsteady flow equation*, in *Unsteady Flow in Open Channels*, 1, Edited by Mahmood,K. and Yevjevich,V., Water Resources Publ., Fort Collins, Colorado

Masood Ul-Hassan, K. and Wilson, G. (1995). *Hydroinformatic Applications in Real Time Control Strategy Selection,* in *HYDRA 2000* (Vol. 5) Thomas Telford, London

Minns, A.W. and Babović, V. (1996). *Hydrological modelling in a hydroinformatics context*, in Abbott, M.B., and Refsgaard, J.C., (Editors), *Distributed Hydrological Modelling*, Chapter 14, Kluwer Academic Publishers, Dordrecht, The Netherlands

Nujić, M. (1995). *Efficient implementation of non-oscillatory schemes for the computation of free-surface flows,* Journal of Hydraulic Research, Vol. 33, No. 1, pp. 101-111

Oden, J.T. (1989). *Progress in Adaptive Methods in Computational Fluid Dynamics,* pp.206-252, in Flaherty, J.E., Paslow, P.J., Shephard,M.S. and Vasilakis, J.D. (Editors), *Adaptive Methods for Partial Differential Equations*, SIAM, Philadelphia

Oliger, (1984). *Adaptive Grid Methods for Hyperbolic Partial Differential Equations*, pp.320-331, in Santosa, F. Symes, W.W. and Holland, C. (Editors), *Inverse Problems of Acoustic and Elastic Waves*, SIAM, Philadelphia

Osiadacz, A.J. (1987). *Simulation and Analysis of Gas Networks,* Spon, London

Parter, S.V. (1961). *The use of linear graphs in Gaussian elimination*, SIAM Review 3, pp. 119-130

Pissanetsky, S. (1984). *Sparse Matrix Technology*, Academic Press Inc.,London

Rahman, M. and Chaudry, M.H. (1995). *Simulation of hydraulic jump with grid adaptation*, Journal of Hydraulic Research, Vol. 33, No. 4, pp. 555-567

Rheinboldt, W.C. (1983). *Feedback Systems and adaptivity for Numerical Computations,* in Babuška, I., Chandra, J. and Flaherty, J.E. (Editors), *1. Workshop on adaptive computational methods for partial differential equations*, pp. 3-19, SIAM, Philadelphia

Savic, LJ. and Holly, F.M. (1993). *Dambreak flood waves computed by Godunov method*, Journal of Hydraulic Research, Vol. 31, No. 2, pp. 187-204

Stepelman, R.S. and Winarsky, N.D. (1979). *Adaptive Numerical Differentiation*, Mathematics of Computation, Vol. 33, No. 148, pp. 1257-1264

Stelling, G.S., Wiersma, A.K. and Willemse, J.B.T.M. (1986). *Practical Aspects of Accurate Tidal Computations*, Journal of Hydraulic Engineering, Vol. 112, No. 9, pp. 802-817

Tewarson, R.P. (1971). *Sorting and ordering sparse linear systems,* in Reid, J.K. (Editor) *Large Sparse Sets of Linear Equations*, Academic Press, London

Timoshenko, S. (1955). *Strength of Materials; Part I: Elementary Theory and Problems,* D. Van. Nostrand Company, Inc., pp. 137-165

Tinney, W.F. and Walker, J.W. (1967). *Direct solutions of sparse network equations by optimally ordered triangular factorisation, Proc. IEEE* 55, pp. 1801-1809

Tsujimoto, T. and Kitamura, T. (1990). *Velocity profile of flow in vegetated-bed channels, KHL Progressive report, 1990,* Hydraulic Laboratory, Kanazawa University

Tsujimoto, T., Okada, T. and Omata, A. (1993). *Field Measurement of Turbulent Flow over Vegetation on Flood Plain of River Kakehashi*, *KHL Progressive report, 1993,* Hydraulic Laboratory, Kanazawa University

Vanecek, S., Verwey, A. and Abbott M.B. (1994). *HYPRESS: an exercise in object orientation for water hammer and water distribution simulation in pipe networks*, *Proc. First Int. Conf. Hydroinformatics,* Balkema, Rotterdam

Verboom, G.K., Stelling, G.S. and Officier, M.J. (1992). *Boundary conditions for the shallow water equations*, in Abbott, M.B. and Cunge, J.A. (Editors). *Engineering Applications of Computational Hydraulics*, pp. 230-262, Pitman, London

Verwey, A. (1994). *Linkage of physical and numerical aspects of models applied in environmental studies*, *Proc. Watercomp '94*, Inst.Eng. Australia

Volkov, E.A. (1986). *Numerical Methods*, MIR Publishers, Moscow

Wang, H.H. (1981). *A parallel method for tridiagonal equations,* ACM Trans. Math. Softw. No. 7, pp. 170-183

Watanabe, T. and Kondo, J. (1990). *The influence of Canopy Structure and Density upon the Mixing Length within and above Vegetation*, J.Met. Soc.Japan, Vol. 68, No. 2, pp. 227-235

Weare, T.J. (1979). *Errors arising from irregular boundaries in ADI solutions of shallow-water equations*, International Journal for Numerical Methods in Engineering, Vol. 14, pp. 921-931

Weatherill, N.P. (1990). *Numerical Grid Generation*, Lecture Series 1990-06, von Karman Institute for Fluid Dynamics, Belgium

Weiyan, T. (1992). *Shallow water hydrodynamics*, Elsevier, Amsterdam

Wilders, P., Stijn, Th.L. van, Stelling, G.S. and Fokkema, G.A. (1988). *A fully implicit splitting method for accurate tidal computations*, International Journal for Numerical Methods in Engineering, Vol. 26, pp. 2707-2721

Wilson,G. (1996). *Reinforcement Learning: A New Technique for the Real Time Control of Hydraulic Networks* To appear in *Hydroinformatics 96*, Zurich

Wolfram, S. (1988). *Mathematica: a system for doing mathematics by computer*, Addison-Wesley, Redwood City

Xian, Y.J. (1993). *Quasi-Three-Dimensional Numerical Modelling of Flow and Dispersion in Shallow Water* in *Communications on Hydraulic and Geotechnical Engineering,* Faculty of Civil Engineering, Delft Univ. of Technology, report nr. 93-3

IHE
DELFT

The aim of the International Institute for Infrastructural, Hydraulic and Environmental Engineering, IHE Delft, is the development and transfer of scientific knowledge and technological know-how in the fields of transport, water and the environment.

Therefore, IHE organizes regular 12 and 18 month postgraduate courses which lead to a Masters Degree. IHE also has a PhD-programme based on research, which can be executed partly in the home country. Moreover, IHE organizes short tailor-made and regular non-degree courses in The Netherlands as well as abroad, and takes part in projects in various countries to develop local educational training and research facilities.

International Institute for
Infrastructural, Hydraulic and
Environmental Engineering

P.O. Box 3015
2601 DA Delft
The Netherlands

Tel.: +31 15 2151715
Fax: +31 15 2122921
E-mail: ihe@ihe.nl
Internet: http://www.ihe.nl

For Product Safety Concerns and Information please contact our EU representative GPSR@taylorandfrancis.com
Taylor & Francis Verlag GmbH, Kaufingerstraße 24, 80331 München, Germany

www.ingramcontent.com/pod-product-compliance
Lightning Source LLC
LaVergne TN
LVHW080653011026
849160LV00014B/1405
* 9 7 8 9 0 5 4 1 0 4 0 5 6 *